RACE TO SAVE THE PLANET

STUDY GUIDE
2001 EDITION

Edward C. Wolf

Part of a college-level telecourse that includes
Living in the Environment:
Principles, Connections, and Solutions or
Environmental Science: Working with the Earth
written by G. Tyler Miller, Jr., and published
by Brooks/Cole Publishing Company

Produced by WGBH-TV, Boston

Brooks/Cole
Thomson Learning.™

Australia • Canada • Mexico • Singapore • Spain • United Kingdom • United States

Cover Image Credits:
Middle: Protest. © G. Brad Lewis/Getty One

Part opening image:
The earth from space. NASA

Cover Design: WGBH Design

*This book is printed on
acid-free recycled paper.*

BROOKS/COLE
A Division of Thomson Learning
The Thomson Learning logo is a registered trademark
under license.
For more information, contact BROOKS/COLE,
511 Forest Lodge Road, Pacific Grove, CA 93950, USA
or electronically at http://www.brookscole.com

Printed in the United States of America

1 2 3 4 5 6 7 8 9 10

ISBN 0-534-37807-2

This book was developed for use by students enrolled in
the *Race to Save the Planet* telecourse. This thirteen-unit
telecourse consists of ten one-hour public television
programs, this Study Guide, one of two introductory
environmental science textbooks, and additional
supplementary readings.

Race to Save the Planet is produced for PBS by the
WGBH Science Unit in association with The Chedd-
Angier Production Company, Film Australia, and the
University Grants Commission of India/EMRC/Gujarat
University.

Major funding is provided by the Annenberg/CPB
Project. Corporate funding is provided by Ocean Spray
Cranberries, Inc. Additional funding is provided by the
Jessie Smith Noyes Foundation, the John D. and
Catherine T. MacArthur Foundation, the Carnegie
Corporation of New York, the Geraldine R. Dodge
Foundation, the Hitachi Foundation, the W. Alton Jones
Foundation, the Joyce Foundation, the Charles Stewart
Mott Foundation, the Estate of Laura Atherton
Lawrence, the Andrew W. Mellon Foundation, the
Public Welfare Foundation, the Rockefeller Foundation,
the Larsen Fund, the Rockefeller Brothers Fund, the
Corporation for Public Broadcasting, and public
television viewers.

Funding is also provided by the Marilyn Simpson
Charitable Trust, the Glen Eagles Foundation, the Town
Creek Foundation, United Nations Fund for Population
Activities, UNICEF, NORAD, and the Wallace Genetic
Foundation.

The series concept is based on the Worldwatch
Institute's *State of the World* reports. *Race to Save the
Planet* is closed captioned for hearing-impaired viewers
by The Caption Center at WGBH.

For more information about telecourse licenses and
off-air taping, contact:

Race to Save the Planet
PBS Adult Learning Service
1320 Braddock Place
Alexandria, VA 22314-1698
1-800-ALS-ALS-8

For information about *Race to Save the Planet*
videocassettes and print materials, off-air taping and
duplication licenses, and other video and audio series
from the Annenberg/CPB Collection, contact:

Race to Save the Planet
The Annenberg/CPB Project
P.O. Box 2284
South Burlington, VT 05407-2284
1-800-LEARNER

ACKNOWLEDGMENTS

The distinguished members of our Scientific Advisory Board and of our Utilization Advisory Board, listed below, provided invaluable assistance in developing both the programs and the print materials for this telecourse.

Scientific Advisory Board

Edward Ayensu
International Union of Biological Sciences

David Bell
Harvard University

Lester R. Brown
Worldwatch Institute (ex officio)

Anthony Cortese
Tufts University

Harold Dregne
Texas Tech University

Charles Ebinger
Georgetown University

John Holdren
University of California, Berkeley

Walter Jackson
USX Corporation

Donald Lesh
Global Tomorrow Coalition

Joseph Ling
3M Corporation (retired)

Donella Meadows
Dartmouth College

David Pimentel
Cornell University

Peter Raven
Missouri Botanical Garden

Amulya K. N. Reddy
Indian Institute of Science

Roger Revelle
University of California, San Diego

William Ruckelshaus
Browning-Ferris Industries

Stephen Schneider
National Center for Atmospheric Research

Carl Weinberg
Pacific Gas and Electric Company

George Woodwell
The Woods Hole Research Center

Utilization Advisory Board

Dennis Gathmann
Lake Land College

Colin High
Dartmouth College

James Hornig
Dartmouth College

Donella Meadows
Dartmouth College

William R. Moomaw
Tufts University

John Wargo
Yale University

Susan U. Wilson
Miami-Dade Community College

Thomas Wilson
Open University of South Florida

WGBH Educational Foundation
Boston, Massachusetts

Production Team

Peter McGhee
Manager
National Program Productions

Paula Apsell
Director
Science Unit

John Angier
Executive Producer

Linda Harrar
Senior Producer

Print Development

Brigid Sullivan
Manager
Special Telecommunications Services

Ann Strunk
Director
Print Projects

Leah Osterman
Coordinator
Print Projects

Patricia Crotty
Coordinator
Print Projects

CONTENTS

Preface: The *Race to Save the Planet* Telecourse

Course Overview

When roaming bands of hunting-gathering peoples first settled in permanent communities in the Middle East about 12,000 years ago, there were no more than a few million human beings on earth. Today, there are 6 *billion* of us, and our numbers increase by 80 million each year. That extraordinary growth, a sign of our success as a species, has been both cause and effect of profound revolutions in the relationship between humanity and the earth.

In recent years, the collective impacts of *Homo sapiens* have come under increasing scrutiny as evidence of fundamental changes in the natural patterns and processes of the earth becomes unmistakable. Changes of the sort associated with the passing of geological eons now unfold within human lifetimes. The industrial, agricultural, and energy technologies that sustain our societies have environmental impacts whose consequences can no longer be ignored. The contamination of air and water, extinction of plant and animal species, loss of productive soil, accumulation of waste, and change in the composition of the atmosphere itself lend special urgency to the task of responding to environmental problems. The environment has become a regular topic on networks and in newspapers, and environmental concerns touch every community around the world. Informed citizenship now requires a grasp of environmental issues.

Race to Save the Planet, an introductory course in environmental science consisting of ten one-hour programs, an introductory environmental science textbook, supplementary readings, and this Study Guide, will help you understand how scientists investigate global environmental changes, why disagreement exists about responses to those changes, and how different communities and societies address environmental challenges. Each program is designed to stand alone as well as to fit into the overall focus and scope of the course, which begins with a look at the long sweep of human history and concludes with a summary of the opportunities and challenges with which societies will contend beyond the year 2000.

Race to Save the Planet visited 29 countries around the world to build a global perspective on environmental change. The course introduces fundamental concepts of matter, energy, and ecology, then applies these concepts to a range of contemporary concerns—from the landfill crisis in the United States to the destruction of tropical rain forests. You will see how different societies have responded to common problems and how experimental solutions are being tested. You will gain an appreciation for the complexity of environmental problems such as global climate change and the sophistication of the scientific tools being used to understand them. The course components are designed to work together to enrich your understanding of contemporary environmental concerns, innovative responses, and hard choices for the future—and to emphasize the solutions that can begin in your own home, college, and workplace.

Course Objectives

The *Race to Save the Planet* telecourse is designed to help you

- understand how human impacts on earth have changed through history and why environmental concerns have recently become so prominent.

- recognize the major environmental challenges facing modern societies and understand the choices and trade-offs these challenges pose.

- grasp the scientific principles underlying basic phenomena of environmental change.

- understand the technologies associated with major environmental problems and the technologies that may help solve these problems.

- distinguish the environmental impacts of industrial and developing societies, and understand why different types of societies perceive different problems and pursue different solutions.

- broaden your familiarity with world geography and international affairs.

- understand how the issues discussed in the telecourse are connected to the decisions and choices you make in your personal life.

Course Components

Race to Save the Planet consists of the following components:

- Ten one-hour television programs

- This Study Guide

- One of two textbooks, chosen by the instructor: *Living in the Environment* (11th edition, 2000) or *Environmental Science: An Introduction* (8th edition, 2001), both written by G. Tyler Miller, Jr., and published by Brooks/Cole Publishing Company

- Supplementary texts, to be assigned by the instructor: *State of the World 2000,* by Lester R. Brown et al., published by W. W. Norton & Company, Inc., and *Taking Sides: Clashing Views on Controversial Environmental Issues,* edited by Theodore D. Goldfarb and published by The Dushkin Publishing Group, Inc.

Together, the course components introduce broad concepts and scientific terms fundamental to ecology; they also examine major environmental issues through a variety of case studies.

The Television Programs

The ten one-hour television programs, corresponding with Units 3 through 12 of the telecourse, explore important ecological concepts and how different communities and societies in 29 countries around the world address environmental challenges. The programs include docu-mentary footage, interviews, simulation, computer graphics, and animation.

The Study Guide

Ten of the thirteen Study Guide units correspond directly to the ten television programs, and three additional print-only units complete the course. Two print-only units precede the television programs, and a final print-only unit concludes the course. The Study Guide is designed to prepare you to view each television program critically and to help you evaluate your understanding of what you have seen. It also integrates the program material with the textbook and reinforces themes that unify the telecourse.

The Study Guide is not a substitute for the assigned and supplementary readings and may not cover all the material your instructor will expect you to master in each unit. Be sure you understand clearly what additional material, if any, you are responsible for in each unit. Use the Study Guide as a tool to organize your study and aid your comprehension of the television programs, not as a crutch to support you through the course.

Each unit of the Study Guide includes four major sections:

1. **Before You View the Television Program**

- **Learning Objectives**—a list of six to ten objectives that highlight concepts and content you should be sure to master.

- **Reading Assignment**—chapter(s) from the textbook your instructor has selected to use with the course.

- **Unit Overview**—a short summary of the television program and reading assignment, emphasizing major themes of the unit.

- **Glossary of Key Terms and Concepts**—a selection of terms and concepts helpful as background before viewing the television programs.

2. **After You View the Television Program**

- **Consider What You Have Seen**—a discussion of the themes and highlights of the corresponding television program, designed to help you see connections within each unit and among units.

- **Take a Closer Look at the Featured Countries**—a brief profile of each of the countries included in the program, with special emphasis on the unit topic, accompanied by maps.

- **Examine Your Views and Values**—several open-ended questions to promote critical thinking and reflection and to help you relate material from the unit to your own life.

3. **Test Your Comprehension**

- **Self-Test Questions**—ten to fifteen multiple-choice and short-answer questions based on the content of the television program, with answers listed in the Appendix.

- **Sample Essay Questions**—five essay questions like those your instructor may assign on exams that will require you to apply concepts learned in the reading assignment to cases and examples from the accompanying television program.

4. **Get Involved**

- **References**—a short bibliography of books and articles that you can consult for further information on the unit topic.

- **Organizations**—an illustrative list of nonprofit and independent organizations involved with the issues featured in the unit.

The Textbook

This Study Guide has been developed for use with either of two textbooks by G. Tyler Miller, Jr.: *Living in the Environment* (11th edition, 2000) or *Environmental Science: An Introduction* (8th edition, 2001), both published by Brooks/Cole Publishing Company. Either of these texts can serve as an integral part of the telecourse. The assigned textbook readings for each unit present fundamental concepts and principles illustrated by the television programs and provide definitions and illustrations that clarify those concepts. Your

instructor will test your comprehension of both the programs and the textbook readings.

You may find that the television programs have a somewhat different emphasis than the text material. The Study Guide will help you link the programs with the textbook.

The Supplementary Texts

Your instructor may choose to assign either or both of the following texts to reinforce the material presented in the television programs and the primary text.

- *State of the World 2000*, by Lester R. Brown et al. (New York: W. W. Norton & Company, Inc., 2000), is the latest in a series of annual reports from Worldwatch Institute that inspired the development of the *Race to Save the Planet* television series. Subtitled "A Worldwatch Institute Report on Progress Toward a Sustainable Society," *State of the World* provides an up-to-date and thoroughly documented overview of the earth's ecological health.

- *Taking Sides: Clashing Views on Controversial Environmental Issues*, edited by Theodore D. Goldfarb (Guilford, Conn.: The Dushkin Publishing Group, Inc., 1997), is a collection of essays providing opposing viewpoints on a number of contemporary environmental issues. Many of the issues in *Taking Sides: Clashing Views on Controversial Environmental Issues* bear directly on material in *Race to Save the Planet*, and the exploration of antithetical views will help you sharpen your critical thinking skills. Taking Sides ® is a registered trademark of The Dushkin Publishing Group, Inc.

If neither of these supplementary texts is assigned by your instructor, you may wish to purchase them (or check them out from a library) and read them on your own as you take the telecourse. Suggested readings from both books are included in the References section at the end of each Study Guide unit.

To order copies of *State of the World 2000*, contact:

W. W. Norton & Company, Inc.
Attn: Laurie Garner
500 Fifth Avenue
New York, NY 10110
(212) 354-5500

To order copies of *Taking Sides: Clashing Views on Controversial Environmental Issues,* contact:

The Dushkin Publishing Group, Inc.
Sluice Dock
Guilford, CT 06437
(203) 453-4351
(800) 243-6532

Taking the Race to Save the Planet Telecourse

Find out the following information as soon after registration as possible.

- What books are required for the course.

- If and when an orientation session has been scheduled.

- When *Race to Save the Planet* will be broadcast in your area.

- When course examinations are scheduled (mark these on your calendar).

- If any additional on-campus meetings have been scheduled (plan to attend as many review sessions, seminars, and other meetings as possible).

To learn the most from each unit:

1. Before viewing the television program, read the "Before You View the Television Program" section of the corresponding unit in the Study Guide, paying particular attention to the Learning Objectives and the Glossary of Key Terms and Concepts.

2. Read the textbook assignment listed in the Study Guide and any supplementary material assigned by your instructor. Chapter outlines, lists of key concepts, and summaries will help you identify the important information.

3. View the program, keeping the Learning Objectives in mind. Be an active viewer. Some students find that taking notes while viewing the programs is helpful. If your area has more than one public television station, there may be several opportunities for you to watch the program. Many public television stations repeat a program at least once during the week it is first shown. The films may also be available on videocassettes at your school, or you can tape them at home if you own a VCR. If you don't have a VCR, you can make an audiocassette of the program for review.

4. Read the "After You View the Television Program" section of the Study Guide, to review and expand on what you have seen.

5. In the "Test Your Comprehension" section, take the self-test questions and complete any other questions, activities, or essays assigned by your instructor.

6. Keep up with the course on a weekly basis. Each unit of the course builds upon knowledge gained in previous units. Stay current with the programs and readings. Make a daily checklist and keep weekly and term calendars, noting your scheduled activities such as meetings or examinations as well as blocks of time for viewing programs, reading, and completing assignments.

7. Keep in touch with your instructor. If possible, get to know him or her. You should have your instructor's mailing address, phone number, and call-in hours. Your instructor would like to hear from you and to know how you are doing. He or she will be eager to clarify any questions you have about the course.

RACE TO SAVE
THE PLANET

PART I

THE AGE OF
GLOBAL CHANGE

UNIT 1

The Age of
Global Change

The Naam movement in Sub-Saharan Africa fosters local leadership and community work during the long dry season to promote more abundant crops and wipe out hunger —"faim"—in the region.

Can we move nations and people in the direction of sustainability? Such a move would be a modification of society comparable in scale to only two other changes: the agricultural revolution of the late Neolithic and the Industrial Revolution of the past two centuries. These revolutions were gradual, spontaneous, and largely unconscious. This one will have to be a fully conscious operation, guided by the best foresight that science can provide—foresight pushed to its limit. If we actually do it, the undertaking will be absolutely unique in humanity's stay on earth.

    William D. Ruckelshaus, "Toward a Sustainable World." Scientific American, Sept. 1989.

GETTING STARTED

Learning Objectives

After completing the assigned readings, you should be able to

- explain the principle of exponential growth and cite two real-world cases that demonstrate this type of growth.

- explain the distinction between "renewable" and "nonrenewable" resources and list examples of each.

- describe the two major categories of pollution and the two principal approaches to pollution control.

- paraphrase and explain the implications of the law of conservation of matter.

- paraphrase and explain the implications of the first law of thermodynamics, the "law of conservation of energy."

- paraphrase and explain the implications of the second law of thermodynamics, the "law of energy quality degradation."

- define energy efficiency and net useful energy.

- explain how typical uses of energy in daily life demonstrate the two laws of energy.

- list some of the characteristics of what textbook author Miller calls "sustainable-earth societies."

- summarize your own worldview and say whether it more closely resembles the "throwaway" or the "sustainable-earth" worldview discussed in the reading assignment.

Reading Assignment

In this unit you may read the Study Guide and the text reading assignment in any order you prefer. You will have to complete the textbook assignment before taking the self-test questions at the end of the unit, however. Unit 2, like this unit, provides space in which to fill in your own definitions of fundamental concepts after you complete the reading assignment. In Units 3 through 12, schedule time to complete the assigned reading and the first section of the Study Guide ("Before You View the Television Program") *before* the accompanying program is broadcast or shown in class.

    Choose the material from either textbook as your reading assignment. Your instructor might assign additional readings as well.

Living in the Environment

Chapter 1, "Environmental Problems, Their Causes, and Sustainability" (esp. Section 1-7)
Chapter 2, "Critical Thinking: Science, Models, and Systems" (esp. Section 2-1)
Chapter 3, "Matter and Energy Resources: Types and Concepts"
Chapter 29, "Environmental Worldviews, Ethics, and Sustainability"

Environmental Science

Chapter 1, "Environmental Problems and Their Causes" (esp. Section 1-7)
Chapter 2, "Economics, Politics, Ethics, and Sustainability" (esp. Sections 2-7 and 2-8)
Chapter 3, "Science, Systems, Matter, and Energy"

    The references listed at the end of this unit may help you begin thinking about the earth and the future in a new way and begin to develop your own perspective on the age of global change.

Glossary of Key Terms and Concepts

The glossary section of Units 3 through 12 in this Study Guide includes terms and concepts it will be helpful for you to understand before watching the accompanying television program. Many of these terms are not defined in the reading assignments; you should get acquainted with them before viewing the programs so that you don't need to spend time flipping through the Study Guide while you watch.

The reading assignment for this unit introduces fundamental principles that you will use throughout this course. Be sure that you have a thorough grasp of the following concepts before you begin the units accompanied by television programs. You will find complete definitions and explanations in your textbook, and you might find it helpful to paraphrase the explanations in your own words in the space provided.

---

1. The Dynamics of Growth

exponential growth (and its consequences)

2. Types of Resources

renewable resources

nonrenewable (exhaustible) resources

sustainable yield

3. Types of Pollution, Responses to Pollution

point sources of pollution

nonpoint sources of pollution

input pollution control

output pollution control

4. Basic Principles of Matter and Energy

law of conservation of matter

first law of thermodynamics

second law of thermodynamics

energy quality

energy efficiency

net useful energy

5. Ways of Looking at the Earth and Society

"high-throughput" worldview

"low-throughput" worldview

TAKE A LOOK AHEAD

The Television Course at a Glance

Race to Save the Planet provides an up-to-date look at the global environmental challenges confronting humanity and the upsurge in environmental awareness that is sweeping societies around the world. This thirteen-unit telecourse consists of ten television programs with accompanying print units and three print-only units, produced by WGBH-TV in Boston with support from the Annenberg/CPB Project.

To relate the ten television programs more clearly to the objectives of an introductory course in environmental science, the course is divided into three major parts:

Part I.   The Age of Global Change
(Units 1–6)

Part II.   Making Choices for the Future
(Units 7–10)

Part III.   Steps Toward Sustainability
(Units 11–13)

The first part, "The Age of Global Change," seeks to understand human impact on the earth in its broadest terms and considers two categories of societies—affluent, industrial societies and impoverished, developing societies—in which the prevailing ways of meeting human needs today are damaging the systems and processes that support life on earth. This section includes the first four programs in the series:

The Age of Global Change (Unit 1), a print-only unit, provides an overview of the telecourse, introduces basic concepts of matter and energy, and invites students to begin to examine the environmental impact of their individual lives.

Understanding Ecosystems (Unit 2) is a print-only unit that presents the fundamentals of the science of ecology. The assigned readings discuss the cycling of matter and flow of energy through ecosystems and describe the major types of terrestrial and aquatic ecosystems and the physical factors that shape them.

The Environmental Revolution (Unit 3) examines the two major revolutions that have fundamentally changed how human societies use resources, the agricultural revolution and the Industrial Revolution, and suggests that a *third* revolution—one in global awareness of human impacts—is beginning to cause equally far-reaching changes in society as evidence of global-scale environmental damage becomes too obvious to ignore.

Only One Atmosphere (Unit 4) looks at the new understanding of how human activities are altering the earth's atmosphere. The program emphasizes damage to the ozone layer and the greenhouse effect, explores the probable consequences of these changes, and considers how societies can respond.

Do We Really Want to Live This Way? (Unit 5) profiles the pollution problems associated with the world's most affluent societies, including case studies of air quality in the Los Angeles basin in southern California and water quality in the Rhine River basin of northern Europe.

In the Name of Progress (Unit 6) looks at the environmental price tag of conventional industrial development in Third World countries, examines population growth and its relationship to levels of economic development, and profiles community-based "sustainable development" alternatives in Brazil and India.

Part II, "Making Choices for the Future," focuses on four key dimensions of human activity that have implications for the global environment and profiles efforts to choose technologies and approaches that will be less environmentally damaging and more sustainable than present methods.

Remnants of Eden (Unit 7) considers human relationships with ecosystems and the diversity of plant and animal species they support. The program explains present concern about widespread extinctions of species and profiles successful efforts to protect and restore biological diversity by involving local communities in decision making and attempts to balance economic activities with ecological integrity.

More for Less (Unit 8) looks at human uses of energy resources and how heavy reliance on fossil fuels (coal, oil, natural gas) has put the earth's climate at risk. The program explores alternatives to present energy technologies and looks at communities that have taken steps to meet their energy needs through improved efficiency and reliance on renewable energy sources.

Save the Earth—Feed the World (Unit 9) looks at world food production and the challenge of feeding a growing population. The program discusses the environmental damage caused by conventional farming in both industrial and developing countries and the worldwide movement to replace conventional methods with low-input, sustainable agricultural practices.

Waste Not, Want Not (Unit 10) examines the mounting problem of waste disposal, comparing different countries' methods of managing solid waste, toxic waste, and sewage. Methods to reduce the volume of trash generated and to dispose of waste materials in more sustainable, less damaging ways are featured.

The final part, "Steps Toward Sustainability," concerns how humanity will make the adjustments required to bring human needs and wants in balance with environmental limits as evidence of global changes accumulates. The two programs in Part III look at responses to global environmental problems emerging at the individual, national, and international levels.

It Needs Political Decisions (Unit 11) profiles three countries that have chosen to address the challenge of environmental sustainability at the *national* level: Zimbabwe, Thailand, and Sweden. This program looks at the trade-offs and difficult choices made within each society, progress made to date, and the most serious problems that confront national leaders.

Now or Never (Unit 12) presents the search for sustainability at the *individual* and *international* levels—by profiling individuals whose courage and imagination have produced innovative responses to global problems and highlighting international initiatives designed to address the challenge of global change. The program examines how new environmental problems are causing nations to redefine their concept of security and changing the world's political geography.

Beyond the Year 2000 (Unit 13), the final print-only unit, reviews the challenges and solutions presented in the ten television programs and considers the likelihood that the fundamental adjustments needed to achieve sustainability will be made by the turn of the century.

Unifying Themes

Along with the terms, concepts, and analytic tools that will be introduced in each unit of *Race to Save the Planet* television course, five major themes will carry through the entire course. Although they may not always be discussed explicitly, you should look for instances of each theme in the ten television programs. By the end of the course, you will be able to find abundant evidence of these themes in the world around you.

The Rate of Change

Throughout the television series, you will notice that scientists and other authorities express concern not only about the scope of the problems being discussed but also about the speed at which they are increasing. This increase in the rate of change is one of the most worrisome aspects of many issues introduced in *Race to Save the Planet*. If in each decade a problem grows worse than it has in any preceding decade, it probably exhibits an accelerating rate of change.

Generally, scientists concerned about rate of change are concerned with phenomena that are **growing exponentially**. People are most familiar with linear growth, when a constant amount is added over a given time period. A quantity shows exponential growth when it expands by a *constant percentage of the whole* over a given time period. Another name for it is compound growth—for instance, the growth of funds in an interest-bearing bank account. Natural populations of plants and animals, including the human species, tend to grow exponentially unless their growth is checked by a limited food supply, disease, or a particularly ferocious predator. Economies, if they are growing at all, grow exponentially—and with them, demand for raw materials and energy and production of waste products also grow exponentially.

A useful concept for understanding things growing at compound rates is **doubling time**, which helps us relate exponential growth back to the *absolute* increase it accomplishes. Dividing 70 by the growth rate of a quantity growing exponentially gives *the time in years that it will take that quantity to double in size*. For example, in 1999, the whole human population of the earth, about 6.0 billion people, grew by about 1.33% per year. The doubling time for the human population is 70 divided by 1.33, or 53 years. This means that if the human population continues to grow by 1.33% each year, there will be more than 12 billion people on earth by the year 2052, 52 years from now. Many nations are growing much faster than the world's average, and their doublings will occur much sooner.

Anything that grows exponentially will tend to add a particular absolute quantity in shorter and shorter time periods. Thus, your textbook points out that it took at least 60,000 years for the human population to reach 1 billion people, 123 years to add the second billion, 33 years for the third, 14 years for the fourth, and just 13 for the fifth billion, reached in 1987, and the sixth billion 12 years later in 1999.

As long as human populations and their economies grow exponentially, many of their impacts on the planet also grow exponentially: the clearing of forests, the consumption of oil and coal, the release of gases that cause the greenhouse effect and damage the ozone layer, the production of wastes, and many other topics that will be explored in this course. Things growing exponentially are especially hard to control. This is why the rate of change is of such urgent concern to scientists and others concerned with the fate of the global environment.

Reinforcing Problems

Many of the environmental problems we will consider in this course cannot be treated in isolation; they have multiple effects and are linked to other problems. The coal-burning power plants in the U.S. Midwest that release acid-rain–causing pollutants also release carbon dioxide that contributes to the greenhouse effect. The chlorofluorocarbons (CFCs—discussed in Unit 4) that damage the earth's ozone layer also trap the sun's heat, increasing the rate of global warming. The cutting and burning of tropical forests wipes out large numbers of plant and animal species, adds more carbon dioxide to the atmosphere, and destroys plant life needed to absorb carbon dioxide. These and many other examples show that simple cause and effect is not the way to understand the array of environmental challenges now facing humanity. The problems are mutually reinforcing, interrelated in complex ways.

Reinforcing Solutions

Fortunately, the solutions often have the same multifaceted, mutually reinforcing dimension that makes some environmental problems appear so intractable. Using energy more efficiently strikes a blow at the main cause of both the greenhouse effect and acid rain. Recycling paper, aluminum, and glass saves the energy needed to produce these products from virgin materials, reduces air pollution, and slows the filling of waste dumps. Farming with low-input methods reduces pollution of groundwater from pesticides and chemicals, prevents soil erosion, requires less energy, and lowers farmers' costs, making agriculture more profitable. Shifting to sustainable practices can be difficult for a variety of reasons, but once such practices have been adopted, many innovative approaches show advantages that compound over time. Watch for such reinforcing solutions throughout *Race to Save the Planet.*

The Distributive Dimension

Although many of the problems explored in *Race to Save the Planet* affect all societies on earth, neither responsibility for the causes nor vulnerability to the consequences are equally shared. The world's population is growing by 80 million people each year, but nine of every ten births occur in the developing countries of the Third World. Scientists agree that the rising atmospheric concentration of carbon dioxide will affect the climate of the entire earth, but just one-quarter of the world's population, for the most part the privileged minority in industrial nations, accounts for 70% of the emissions that will cause global warming. Ten children in India are born for each baby born in the United States, but that much smaller increment of Americans uses 3.6 times more energy than the increment in India. Your textbook draws a distinction between "people overpopulation" in developing countries and "consumption overpopulation" in developed countries. It is a distinction we will encounter in many forms throughout the television course.

The Role of the Individual

A final theme that runs throughout the television series and the course is that *individuals matter*. The global challenges that provide our subject are not abstract problems, remote from daily life. They are the collective consequences of choices and decisions made every day by people all over the world. Every decision about what to eat, what to wear, whether to turn a switch on or off how to get from here to there is a decision about how to use resources and how to affect the environment. While it is possible to be blasé about the issues introduced in this course, or to be angry about them, or to be puzzled by them, it is *not* possible to be uninvolved in them. To be a human being is to be part of the "race to save the planet." Understanding how individuals are involved is the key to understanding the problems *and* the solutions discussed in this course.

The Geography of the Environmental Revolution

Race to Save the Planet draws on examples and case studies from all over the world, and gaining a better understanding of the geography of environmental problems is an important goal of the course. Each unit in the Study Guide includes maps and short profiles of countries featured in the accompanying television program, providing a country-specific perspective on the unit's theme.

Figure 1-1 shows a map of the world on which the countries featured in the television series are highlighted. All countries have a common stake in successfully addressing global challenges such as the greenhouse effect or deforestation, even though responsibility for the causes is not shared equally. By the end of the course, you will have a clearer sense of the economic and social divisions among societies, a knowledge of the environmental problems that confront countries at different latitudes and different stages of economic development, and a feeling for the leadership that has arisen in industrial and developing countries alike as the environmental revolution has spread around the world.

TAKE STOCK OF YOUR LIFESTYLE

Another primary goal of this course is to help you realize how inextricably your own life is bound in the systems and resources of the global environment, how you share responsibility for the changes taking place, and how you can use personal choices in your own household to address global environmental challenges, whether or not you are involved in an environmental profession.

While the course will expose you to new facts and concepts, it should also enable you to look at your own life in an entirely new way. During this first week of the course, before you have begun to watch the television series, take a close look at the resources you depend on and the way you use those resources. As you begin the course, develop a short profile of your lifestyle that will provide a personal point of departure for the issues that will be covered. No specialized knowledge is needed to do this, just a willingness to look at some parts of

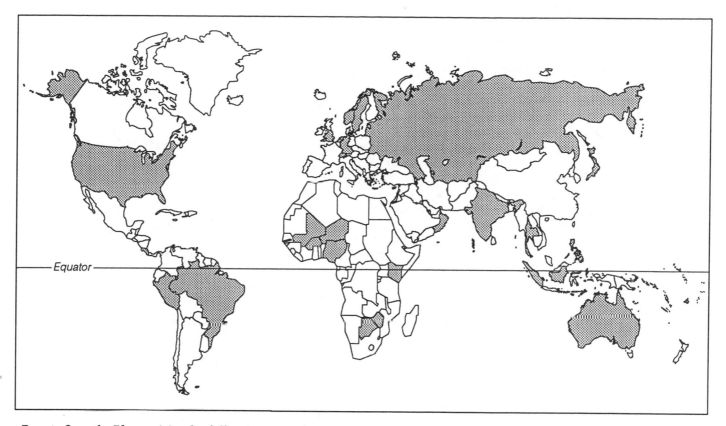

Race to Save the Planet visits the following countries:

Australia*	Denmark*	Jordan	Norway	Sweden*
Botswana	England	Kenya*	Oman*	Switzerland
Brazil*	India*	Mali	Peru*	Thailand*
Burkino Faso	Indonesia*	The Netherlands*	Philipines	United States
Costa Rica*	Israel	Niger	Former Union of Soviet	Former West Germany
Dijbouti*	Japan*	Nigeria	Socialist Republics*	Zimbabwe

•Featured in Study Guide country profiles

Figure 1-1

your life more closely than you have before. You can add to or change this personal inventory as you go through the course.

Include in your profile a short paragraph that summarizes what you know about each of the following topics. You probably can provide most of this information off the top of your head. In some cases, a phone call or two to local officials should be enough to fill in missing details.

1. Basic Resources

Food   Take a look at your diet. What proportion of your diet is animal foods? What proportion is vegetable foods? How much of your food is grown locally? How much (approximately) is grown in the United States? How much is imported?

Energy   List the forms of energy used in your household. What are the main uses of each type (e.g., electricity for lighting and appliances, gas for cooking and heating water)? Make a list of the major energy-using appliances you own or use each day.

Transport   How do you get from place to place? How much of your travel is done by automobile? How much by bus or other mass transit? What is the approximate fuel economy of the automobile you use? How often do you walk or use a bicycle for a nonrecreational purpose?

Waste   What are the major components of your household trash? Do you separate any materials from your trash for recycling? Where does your household trash go? How and where is sewage from your household treated (for example, home septic system, town sewage treatment plant)?

2. The Wider Community

Population   How large is your community? Is it growing or shrinking in size? How many people live in your state? Is it growing or shrinking? Find and include national population figures for purposes of comparison.

The Local Ecosystem   Describe the physical appearance of your area. Is it forested? Dominated by deciduous or evergreen trees? Is it prairie? Desert? Are any major rivers or lakes nearby? How much of your area remains in a "natural" state? Can you name any characteristic plant or animal species?

Environmental Problems   Make a list of pollution or other environmental problems in your local community. These could include anything from gypsy moth infestation to an abandoned industrial dump site. Describe any local efforts (laws, volunteer efforts, etc.) to respond to these problems. Are there any major problems that are not currently being addressed?

Environmental Hazards   What major risks is your community vulnerable to? Do you live in a coastal area or a flood plain? An area prone to earthquakes, droughts, or severe storms? Near a power station or major industrial facility?

3. Awareness

The Present   While you are enrolled in this course, keep track of environmental news in a daily local or national newspaper or a national newsmagazine. If a particular topic, such as the greenhouse effect or the landfill crisis, interests you, clip and save articles dealing with this topic. Save your newspapers for recycling, and write an essay or make a class presentation based on the topic you have followed during the course.

The Future   What are your aspirations and expectations about the future? What problems do you believe *have* to be solved during your lifetime? If you have or plan to have children, do you think they will live the way you live today?

In an article titled "Managing Planet Earth" listed in the references section at the end of this unit, environmental scientist William Clark observes: "What kind of planet we want is ultimately a question of values." By taking this course, you will gain deeper insight into how human activities affect planet earth and, even more importantly, how the values reflected in your personal lifestyle—in your choices about what to buy, how to travel, and how to manage your household—have a direct bearing on our collective choice about the "kind of planet we want."

TEST YOUR COMPREHENSION

Self-Test Questions
(*Answers at end of Study Guide*)

Multiple Choice

1. The J-curve of human population discussed in the reading assignment is a graphic representation of
    a.   linear growth.
    b.   the human birth rate.
    c.   constant growth.
    d.   exponential growth.

2. Which of the following is *not* an example of exponential growth?
    a.   the expansion of the human population
    b.   a 5 1/4% savings account

c. the increase in carbon dioxide emissions since 1850

d. the production of automobiles by a single factory

e. the production of automobiles worldwide

3. Harvesting more than the sustainable yield of a renewable resource
   a. leads to environmental degradation and makes the resource nonrenewable.
   b. is a contradiction in terms.
   c. is an appropriate way to use resources in less developed countries.
   d. causes exponential growth that regenerates the resource.
   e. is an appropriate way to use resources in more developed countries.

4. The two general types of pollution control are
   a. renewable and nonrenewable methods.
   b. perpetual and exhaustible methods.
   c. input and output methods.
   d. point and nonpoint methods.
   e. linear and exponential methods.

5. "High-quality" forms of energy are
   a. forms that cause no air pollution.
   b. concentrated forms capable of performing work.
   c. forms used in more developed countries.
   d. often very dilute.
   e. produced by converting matter into energy.

6. The first law of energy
   a. applies to all forms of energy except nuclear power.
   b. applies to all forms of energy except solar power.
   c. says that energy input always equals energy output.
   d. tells what products to expect in a nuclear reaction.
   e. says that energy quality and energy quantity are interchangeable.

7. The law of conservation of matter
   a. is known as the first law of thermodynamics.
   b. means that some quantity of waste will always be produced.

c. means that a waste disposal crisis is inevitable.

d. applies to physical processes but not to chemical processes.

e. means that no waste would be produced in an ideal world.

8. The observation that energy quality inevitably declines whenever energy is converted from one form to another
   a. is contradicted by the fact that nuclear power plants generate electricity.
   b. is a demonstration of the law of conservation of energy.
   c. means that future societies must use lower-quality energy sources.
   d. is true of chemical and physical processes but not of living processes.
   e. is a demonstration of the second law of thermodynamics.

9. If societies followed the "sustainable-earth" model discussed in the text,
   a. both materials and energy would be recycled efficiently.
   b. the only waste produced would be the inevitable production of waste heat.
   c. the capacity of the earth to dilute waste and absorb heat would soon be exceeded.
   d. the only form of energy used would be nonpolluting solar energy.
   e. input approaches would be used to reduce materials waste and prevent pollution.

True or False

1. Ninety-five percent of the energy used in a typical incandescent bulb is lost as heat. _____

2. The most efficient way to heat bath water is to use electricity generated in nuclear power plants. _____

3. Energy efficiency saves energy at more than the cost of generating it. _____

4. Only 16% of the commercial energy used in the United States actually performs useful work. _____

5. Superinsulating a home is the most efficient way to provide heat in a cold climate. _____

6. Net useful energy is the same as the potential energy from a fuel source. _____

Definitions

You should be able to define and supply examples of each of the following terms, introduced and discussed in the reading assignment.

exponential growth

renewable resources

nonrenewable resources

sustainable yield

people overpopulation

consumption overpopulation

energy quality

commercial energy

energy efficiency

net useful energy

entropy

GET INVOLVED

References

A number of books and other publications can help clarify the concept of sustainability that runs throughout this course. To get better acquainted with this worldview, read one or more of the following:

Brown, Lester R., et al. *State of the World 2000.* New York: W. W. Norton & Co., 2000.

Ehrlich, Paul R., and Anne H. Ehrlich. Betrayal of Science and Reason: How Anti-Environmental Rhetoric Threatens Our Future. Covelo, Calif.: Island Press, 1996.

Meadows, Donnella H., et al. *Beyond the Limits: Confronting Global Collapse, Envisioning a Sustainable Future.* White River Junction, VT: Chelsea Green, 1992.

Myers, Norman, ed. Gaia: An Atlas of Planet Management. Garden City, NY.: Anchor/Doubleday. 1995.

Ruckelshaus, William D. "Toward a Sustainable World." *Scientific American,* Sept. 1989.

Simon, Julian. The Ultimate Resource 2. Princeton, N.J.: Princeton University Press. 1996.

World Commission on Environment and Development. *Our Common Future.* Oxford: Oxford University Press, 1987.

Organizations

The issues of global change—climate warming, ozone depletion, the waste disposal crisis, the loss of plant and animal species, deforestation, environmental degradation caused by poverty, and so on—are now part of the programs of every national environmental group and many local community and academic groups. It would be impossible to list them all here; we have tried in later units to include a selection of groups that have a special emphasis on the topic of each unit.

The following organizations are particularly good sources of information on global problems; their research reports supplied much of the factual background used to develop this course. Each can provide you with a list of current publications and a description of their activities, including ways you can get involved in supporting their work.

World Resources Institute
10 G St. NE, Suite 800
Washington, D.C. 20002
Tel: (202) 769-7600; web: www.wri.org/wri

Worldwatch Institute
1776 Massachusetts Avenue NW
Washington, D.C. 20036
Tel: (202) 452-1999; web: www.worldwatch.org

Understanding Ecosystems

ROBERT EBER

The Serengeti Plain, in Kenya and Tanzania, is home to the most diverse group of large mammals remaining on the earth. The Serengeti ecosystem, a grassland with scattered deciduous trees and lengthy dry seasons, supports herds of antelopes and wildebeests, elephants, zebras, giraffes, lions, and cheetahs.

A FEW FUNDAMENTALS

Learning Objectives

After completing the assigned readings, you should be able to

- distinguish the different ecological roles played by producers, consumers, and decomposers in natural ecosystems.

- diagram and describe at least two biogeochemical cycles and discuss the major human impacts on each of those cycles.

- define and discuss the importance of photosynthesis in both terrestrial and aquatic ecosystems.

- explain how the trophic structure of natural ecosystems demonstrates the second law of energy, the law of energy quality degradation.

- list the five factors responsible for the earth's climate and discuss the factor influenced by human activities.

- describe the three main types of terrestrial ecosystems.

- explain how temperature and precipitation levels determine the type of terrestrial ecosystem.

- discuss the two major categories of aquatic ecosystems

- explain the role of salinity in determining the type of aquatic ecosystem.

- list the major distinctions between *temperate* and *tropical* versions of the three types of terrestrial ecosystems.

Reading Assignment

Choose the material from either textbook as your reading assignment. Your instructor might assign additional readings as well.

Living in the Environment

Chapter 4, "Ecology, Ecosystems, and Food Webs"

Chapter 5, "Nutrient Cycles and Soils" Sections 5-1 through 5-6

Chapter 6, "Evolution and Biodiversity: Origins, Niches, and Adaptations"

Chapter 7, "Geographical Ecology, Climate, and Biomes"

Chapter 8, "Aquatic Ecology"

Environmental Science

Chapter 4, "Ecosystems and How They Work"

Chapter 5, "Evolution, Biodiversity, and Community Processes"

Chapter 6, "Climate, Weather, and Biodiversity"

Unit Overview

This print-only unit introduces the fundamentals of the science of ecology, the study of how the plant, animal, and microorganism species in living communities interact with one another and with the chemical and physical factors that make up their nonliving environment. "Ecology" comes from the Greek words *oikos* ("household") and *logos* ("study of"), a reminder that this science is not about some phenomenon "out there" but about *our* home, the planet that sustains *Homo sapiens.*

The first assigned chapter defines ecosystems and considers how they work. Two fundamental processes, common to all ecosystems, are presented: the capture of solar energy and its **flow** through ecosystems (in accordance with the second law of energy) and the retention and **cycling** of nutrients in the biogeochemical cycles. The chapter considers the limiting factors and tolerance ranges that determine the distribution and abundance of individual species of plants and animals and introduces the various ways species coexist within ecosystems. The second assigned chapter describes the basic types of ecosystems found on land and in water and considers how the earth's climate (average temperature and precipitation) affects terrestrial ecosystems and how the concentration of salts dissolved in water (salinity) influences aquatic ecosystems.

Every spot on earth is part of an ecosystem. Every breath that each one of us takes, every meal we eat and drink we sip, ties us directly into the planet's biogeochemical cycles. As we will see in the ten programs of *Race to Save the Planet*, the collective impacts of humankind have now reached a scale at which our activities affect the capture of sunlight and flow of energy through ecosystems, the diversity of plant and animal species that those systems comprise, and the global cycling of gases, mineral elements, and water. Understanding ecosystems and how they work is the first step toward understanding human impacts on the global environment and, further, how we can begin to bring human wants and needs within the boundaries set by the productivity and resilience of the biosphere.

Glossary of Key Terms and Concepts

The key concepts and principles of ecology are defined and illustrated in your reading assignment. However, these concepts are so important to mastery of environmental science and a thorough understanding of the cases you will see in the *Race to Save the Planet* television programs that you should personalize your comprehension. Use the space on the following pages to fill in your own definitions, and refer back to these pages whenever you need a review of the basics later in the course.

Though your instructor may test you on some of these terms, this space is for your own use. Fill it in however you like: Jot down key words, make sketches, write your own definitions with the text closed, or copy directly from the textbook if you prefer. However, you will find that if you make the effort now to put things in your own words, you will understand the concepts better and recognize them more easily in the units ahead.

---

1. **Indispensable Definitions**

   The "big picture" terms:

   biosphere

   ecosystem

   biological community

biome

biogeochemical cycles

The "cogs and wheels" of natural ecosystems and how they turn:
producers

consumers

decomposers

food chain

food web

trophic level

bioamplification

photosynthesis

respiration

net primary productivity

How species coexist within ecosystems:

niche

competition

predation

parasitism

mutualism

commensalism

Physical determinants of ecosystems on land and in water:

climate

salinity

2. Biogeochemical Cycles

Write a short description of the main biogeochemical cycles, in each case summarizing the basic cycle and listing the principal ways that human activities affect it.

carbon cycle

nitrogen cycle

phosphorus cycle

sulfur cycle

hydrological (water) cycle

3. Ecosystem Types

Write a short description of each of the three terrestrial ecosystem types, including how each is determined by precipitation and temperature.

desert

grassland

forest

List the ecosystem types associated with the two categories of aquatic ecosystems and describe their characteristic features.

freshwater ecosystems (discuss "eutrophication" in your response)

saltwater ecosystems

TEST YOUR COMPREHENSION

Self-Test Questions
(Answers at end of Study Guide)

Multiple Choice

1. The proportion of solar energy reaching the earth that is captured by living plants to become the energy source for all ecosystems is
    a. about 34%; the rest is reflected by clouds and snow.
    b. about 66%; the rest is absorbed by the oceans.
    c. between 1% and 10%, depending on the season.
    d. less than 1%; the rest is reflected or absorbed as heat.
    e. nearly 100%; only a tiny fraction is reflected by clouds.

2. The differences among ecosystems on land are mostly due to
    a. differences in average temperature and average rainfall.
    b. differences in component species of plants and animals.
    c. the biogeochemical cycles that distribute nutrients unequally.
    d. variations in the dissolved salts in their water supplies.
    e. the presence or absence of human beings.

3. The process of photosynthesis
    a. converts carbon dioxide and water in the presence of sunlight into glucose and oxygen.
    b. converts glucose and oxygen in the presence of sunlight into carbon dioxide and water.
    c. is carried out by organisms known as producers or autotrophs.
    d. is important to plants but not to other organisms.
    e. More than one of the above is true. The correct answers are _____.

4. The principle that too much or too little of any particular nonliving element of an ecosystem can limit the growth of a species' population is known as the
    a. threshold effect.
    b. limiting factor principle.
    c. range of tolerance.
    d. law of energy quality degradation.
    e. principle of species distribution.

5. In general, observations of energy flow through ecosystems show that
    a. plants capture most of the solar energy available to them.
    b. more trophic levels means a greater loss of high-quality energy from the system.
    c. long food chains capture energy more efficiently than short food chains.
    d. most energy losses could be avoided by more efficient harvesting.
    e. high-quality energy is only needed for a few ecosystem processes.

6. The net primary productivity of an ecosystem is
   a. the difference between the energy captured by plants and the energy used by animals.
   b. the amount of energy retained at the top of the ecosystem's food chain.
   c. the amount by which the energy captured in photosynthesis exceeds the energy used in respiration by plants.
   d. the amount by which the energy produced by green plants exceeds that eaten by consumers.
   e. the number of animal offspring supported by an ecosystem's green plants.

7. The principal human impact on the carbon cycle is
   a. planting crops that are more effective at absorbing carbon dioxide.
   b. burning carbon-containing fossil fuels and destroying forests.
   c. increasing the runoff of nutrients from farmland.
   d. increasing the population and exhaling more carbon dioxide.
   e. reducing the efficiency of photosynthesis.

8. Human-caused changes in the chemical composition of the atmosphere are reason for concern because
   a. the changes are likely to affect the biogeo-chemical cycles.
   b. the changes are likely to affect the earth's temperature.
   c. many plants are finely tuned to a particular atmospheric composition.
   d. such changes led to the demise of past civilizations.
   e. ecosystems cannot adapt to atmospheric changes.

9. In a part of the world where evaporation generally exceeds precipitation, the biome you would be most likely to find is a
   a. tropical rain forest.
   b. deciduous forest.
   c. savanna.
   d. desert.
   e. freshwater wetland.

10. A distinctive feature of the tropical rain forest biome is
    a. most of its nutrients are in the soil rather than in the vegetation.
    b. few of its species have economic value.
    c. it quickly recovers after clearing or disturbance.
    d. most of its nutrients are in the vegetation rather than the soil.
    e. it has many producers and consumers but few decomposers.

11. Which of the following is *not* among the ecosystems with the highest levels of net primary productivity per square meter?
    a. coastal estuaries
    b. tallgrass prairie
    c. tropical rain forest
    d. swamps and marshes

12. The process of eutrophication in lakes is a
    a. natural process that human activities often halt or reverse.
    b. process that only occurs in lakes near farmland.
    c. set of chemical changes that does not affect aquatic life.
    d. change in species composition that does not affect lake chemistry.
    e. natural process that human activities often accelerate.

True or False

1. Major ecosystem types found on land are known as "ecotones." _____

2. A biological community includes the plants, animals, and decomposers found in a particular place, but not the nonliving elements. _____

3. A species includes all populations of a particular organism that can interbreed and produce fertile offspring. _____

4. Organisms called consumers or heterotrophs can manufacture their own food. _____

5. Decomposers assist the recycling of chemical elements within ecosystems. _____

6. Most biogeochemical cycles are too vast to be affected by human activities. _____

7. The carbon cycle involves only the atmosphere and living organisms. _____

8. The carbon cycle involves a linkage between photosynthesis and aerobic respiration. _____

9. The nitrogen cycle is important to living organisms because nitrogen is found in proteins and DNA molecules. _____

10. Nitrogen fixation is carried out by most plants but only some microorganisms. _____

11. Water is transferred from the oceans to land in the hydrologic cycle by the power of the sun. _____

12. The oceans account for a larger share of the earth's total net primary productivity than does any other ecosystem type. _____

13. Farmland has a higher net primary productivity than does any terrestrial ecosystem type except the rain forest. _____

14. Because tropical forests are very productive, they make the best farmland when cleared. _____

15. Harvests can diminish the productivity of any ecosystem by removing nutrients. _____

16. Forests may contain more animal species than do grasslands because their complex structure supplies more distinct niches. _____

Sample Essay Questions

1. Name two ecosystem types in the vicinity where you live or attend school. Describe the energy flow through one of those ecosystems. How many trophic levels can you identify with examples?

2. Use principles from the structure of ecosystems to explain why you would expect grazing animals such as gazelles and antelopes to outnumber predators such as lions on an African savanna.

3. Can the entire earth itself be considered an ecosystem? Use the principles of nutrient cycling, energy flow, and the biotic components of ecosystems to explain your answer.

4. Your text reading discusses how the distribution and abundance of species within ecosystems is determined by limiting factors. Is the human species constrained by a limiting factor? If so, what is it? Explain your answer.

5. Using either the carbon cycle or the nitrogen cycle as an example, explain how human activities can affect a biogeochemical cycle and discuss the probable ecological consequences of the human impacts you describe.

GET INVOLVED

References

Many fine books have been written about ecology. The books listed here have been chosen for their clarity, comprehensiveness, and accessibility to a nonscientific audience. In addition, each of these writers deals not only with the science of ecology but also with the values that underlie the ecologist's way of seeing the world.

Attenborough, David. *The Living Planet.* Waltham, Mass.: Little, Brown & Co., 1985.

Colinvaux, Paul. *Why Big Fierce Animals Are Rare.* Princeton: Princeton University Press, 1978.

Ehrlich, Paul. *Earth.* New York: Franklin Watts, 1987.

Leopold, Aldo. *A Sand County Almanac.* New York: Oxford University Press, 1949.

Whittaker, Robert H. *Communities and Ecosystems.* New York: Macmillan, 1975.

Organizations

The two organizations listed here are the main U.S. professional organizations for scientists concerned with ecology. Both issue publications dealing with current research in ecology and carry out public education activities providing an ecological perspective on contemporary environmental problems.

American Institute of Biological Sciences
7144 I St., Suite 200
Washington, D.C. 20005
Tel: (202) 628–1500
web: www.aibs.org/core/index.html

The Ecological Society of America
9650 Rockville Pike
Bethesda, MD 20814
Tel: (301) 530–7005

The Environmental Revolution

CAROL LYNN DORNBRAND

Bedouin tribesmen demonstrate an ancient technique of making plaster, using wood to burn limestone gathered from a cave in southern Jordan. As early pastoralists began to settle down and practice agriculture, their new preference for wood fires and plaster floors contributed to deforestation of the Middle East.

BEFORE YOU VIEW THE TELEVISION PROGRAM

Learning Objectives

After completing the assigned readings and viewing "The Environmental Revolution," you should be able to

- list and explain the four major factors that affect the size and rate of change of the total human population.

- describe the population dynamics and fertility levels characteristic of hunting and gathering societies, settled agricultural societies, and modern industrial societies.

- compare and contrast the environmental impacts characteristic of each of the three types of societies.

- explain what prompted England's shift to reliance on coal as an energy source and how this shift made the Industrial Revolution possible.

- describe some of the environmental problems associated with early industrialization.

- list two key developments that encouraged the "environmental awakening" of the 1960s.

- discuss the causes and effects of citizen environmental activism in the United States in the 1960s.

- explain the significance of the United Nations Conference on the Human Environment, held in Stockholm, Sweden, in 1972.

Reading Assignment

Choose the material from either textbook as your reading assignment. Your instructor might assign additional readings as well.

Living in the Environment

Chapter 1, "Environmental Problems, Their Causes, and Sustainability" Section 1-6

Chapter 11, "Human Population: Growth, Demography, and Carrying Capacity"

Chapter 23, "Sustaining Ecosystems" Section 23-1

Environmental Science

Chapter 1, "Environmental Problems, Their Causes, and Sustainability," Sections 1-6 and 1-7

Chapter 2, "Economics, Politics, Ethics, and Sustainability," Sections 2-7 and 2-8

Chapter 9, "The Human Population: Growth and Distribution," Sections 9-1 and 9-2

Unit Overview

This unit of *Race to Save the Planet* is about history—the history of human occupancy of planet earth and how two turning points in human history fundamentally changed our species' impact on the earth. The first turning point was the shift from mobile hunting and gathering societies to settled agricultural communities, which first occurred in what is now the Middle East about 12,000 years ago; the second was the harnessing of fossil fuels and subsequent Industrial Revolution, which began in England in the eighteenth century. Each change brought a shift in the factors that determine the size and rate of growth of the human population, leading to the unprecedented population expansion of the twentieth century.

The global environmental changes we have begun to recognize today—changes in the land, waters, and even the composition of the atmosphere—are largely the collective consequences of two centuries of industrialization and of the present needs and aspirations of 6 billion human beings. *Race to Save the Planet* suggests we are poised at the edge of a third, equally far-reaching revolution—a global revolution in awareness of environmental change—leading to a reexamination of the foundations of modern societies and fundamental adjustments in our collective use of the earth and its resources. Both the scale of our impacts today and the rate at which those impacts are growing lend special urgency to this new revolution.

The first assigned text chapter gives a brief overview of the changing impact of *Homo sapiens* on the earth over the past 60,000 years and summarizes the stages through which societies have evolved. The chapter also examines how attitudes toward resources and the environment have changed over the course of U.S. history and looks at the causes and accomplishments of the environmental movement in the three decades since its origins in the early 1960s. The second assigned chapter introduces population dynamics and discusses the factors that determine the size and rate of change of human populations. As you read, you should gain a deeper understanding of the process of population growth, recognize the factors responsible for accelerated growth, and learn how population scientists project future growth and estimate the ultimate size of the human population.

The accompanying television program, "The Environmental Revolution," is the first program of the *Race to Save the Planet* series. It takes a contemporary look at the two great revolutions described above, then provides a brief overview of the television series. The program begins with the vestiges of hunter-gatherer societies that remain, to discover what life may have been like for the long stretch of human history before agricultural settlement. Two archaeological sites in the Middle East, one in northern Israel and the second in Jordan, reveal the details of humanity's uneasy transition to agriculture, a transition that anthropologists now believe was prompted by overuse of the environment. The shift to settled societies also brought the population growth that meant there would be no going back to nomadic hunting and gathering ways.

The program then profiles the Industrial Revolution by visiting its birthplace, England, and examining the shift to the use of coal in the eighteenth century that revolutionized iron smelting, gave rise to factory-based manufacturing, and spurred the growth of cities. An almost simultaneous revolution in agriculture, with new farming methods and scientific breeding of livestock, reinforced the industrial economy. The program examines how England's population responded to this unprecedented change and considers the environmental damage that went hand-in-hand with this new way of exploiting the earth to support society.

Finally, the program considers the global spread of the new industrial model, the worldwide change in populations and resource use patterns that is the backdrop for this course. It examines the origins of the environmental movement in the United States in the 1960s, the first questioning of the costs and impacts of the industrial lifestyle. And it concludes with the United Nations Conference on the Human Environment, held in Stockholm, Sweden, in 1972, that provided the first forum for dialogue and debate among nations about human use of the earth.

Both our understanding of environmental problems and the commitment to tackle them have changed enormously in the 18 years since that conference. In the units ahead, we will look closely at the current understanding of global environmental changes and how nations, communities, and individuals are beginning to transform the human activities that have set those changes in motion.

Glossary of Key Terms and Concepts

The following terms and concepts will be useful as background for viewing "The Environmental Revolution."

The **advanced industrial societies** include the nations of Europe, Japan, and North America whose economies are based on industrial manufacturing and the use of fossil fuels. While virtually all nations have developed an industrial base to some extent, the advanced nations dominate the world economy both in their use of resources and in the total value of their economic activity. The C.I.S. and nations of Eastern Europe, also heavily industrialized, lag behind other industrial nations in technology and innovation.

The **age structure** of a population refers to the percentage of the total population represented by specified age groups. Age structure and fertility levels determine the rate of population growth.

Agricultural productivity generally refers to the amount of a crop harvested from a specified unit of land; changes in farming practices change productivity levels, and with them the number of people a given area of farmland can feed.

The **agricultural revolution** (also known as the "Neolithic Revolution") took place in the Middle East between 10,000 and 12,000 years ago when societies first domesticated wild plants and animals and became dependent on crops and livestock for their food.

Ain Ghazal is an archaeological site near Amman, the contemporary capital of Jordan, where scientists have found evidence that documents the transition from hunting and gathering to farming to livestock-based communities; each of the transitions was prompted when the pattern of resource use on which the previous stage depended became unsustainable.

Chlorinated hydrocarbons, a class of synthetic chemicals first produced in the 1930s, include potent pesticides such as DDT (dichloro-diphenyl-trichloroethane) and other compounds that do not break down in the environment and can be concentrated to poisonous levels in the fatty tissues of fish, birds, and mammals.

The **crude birth rate** is the number of live births per 1,000 people in a given population during a given year.

The **crude death rate** is the number of deaths per 1,000 people in a given population during a given year.

Demography is the scientific study of patterns and change in the human population.

The **demographic transition** is a theory that relates population change to levels of economic development, in which populations shift from a condition characterized by high birth and death rates to one characterized by low birth and death rates in response to improving standards of living. In the middle stage of the transition, when death rates are low but birth rates remain high, populations may expand rapidly for several decades or longer.

The first **Earth Day**, a national "teach-in" on the environment, took place in the United States on April 22, 1970. A milestone of environmental activism, the event involved hundreds of thousands of people around the country and demonstrated public support for a burst of environmental laws and initiatives. The second Earth Day, this time a worldwide event, took place on April 22, 1990.

The **environmental movement** took shape in the United States in the 1960s, when concerned citizens first challenged pollution in court, public interest groups were formed around environmental issues, and public opinion forced the U.S. government to pass laws and policies designed to protect environmental quality. The movement is now truly global in its scope and impact.

Fertility is a demographer's term for the average number of live babies born to each woman in a population during her childbearing years. A population has reached "replacement level fertility" when parents have only enough children to replace themselves, and therefore do not contribute to overall expansion of the population.

Fossil fuels include coal, oil, and natural gas formed in geological deposits from the carbon-rich remains of prehistoric plants, animals, and microbes. These fuels, essentially a highly concentrated form of ancient sunlight, became the primary energy source for human societies during the Industrial Revolution.

Hunting and gathering societies depend entirely on natural ecosystems for their nutritional and resource needs. Generally made up of small, mobile family groups who gather wild plants and hunt wild animals for food, materials, and medicine, they are characterized by relatively low fertility levels and short average life spans. The human species depended exclusively on hunting and gathering for the first 99.9% of its 2-million-year tenure on earth.

The **Industrial Revolution** began in England in the eighteenth century, when the use of coal both in steam engines and for iron smelting enormously increased industrial output. Industrial technologies and the use of fossil fuels quickly spread to other nations, generated unprecedented economic growth, and changed fundamentally the human use of resources and impact on the local and global environment.

The **"limits to growth"** debate was sparked in the early 1970s by a group of young scientists at the Massachusetts Institute of Technology who used computer models to study the interaction of population, technology, and natural resources and concluded that economic growth could not be sustained indefinitely on a planet whose physical resources were finite.

Rev. Thomas **Malthus** (1766–1834) was a British theologian and economist who first argued in *An Essay on Population* (published in 1798) that unless restrained artificially by sexual abstinence or birth control, the natural tendency of populations to grow exponentially would inevitably outstrip food supplies, leading to famine and widespread starvation.

The **Natufians**, a society who lived about 12,000 years ago in what is now northern Israel, are believed to represent the transition from mobile hunting and gathering societies to the sedentary agricultural societies that followed. Archaeologists believe that the Natufians, forced to settle by pressure from neighboring tribes, continued to live by hunting and gathering and disappeared when they exhausted the local environment's ability to support a settled population.

AFTER YOU VIEW THE TELEVISION PROGRAM

Consider What You Have Seen

The following themes may help you relate the issues addressed in "The Environmental Revolution" with the reading assignment.

- Understanding human transitions
- The Industrial Revolution: A closer look
- Attitudes toward the earth

Understanding Human Transitions

"The Environmental Revolution" sets the stage for *Race to Save the Planet* by examining several distinct levels of human impact on the earth and considering the factors responsible for the few major transitions that have punctuated the long course of our history as a species. Each level of human society shown in the program can be characterized by its use of energy and the limits imposed by available energy sources, by distinctive population dynamics, and by its response to scarcity and overuse of fundamental resources. These three themes of energy, population, and scarcity will characterize the examples shown throughout the television series.

Hunting and gathering societies are wholly dependent on the plants and animals of natural ecosystems. Their energy source for food and fuel is the sunlight captured by plants on the landscape. In general, nomadic hunting and gathering societies have no way to accumulate a surplus of either food or fuel energy and survive by an intimate knowledge of the patterns of productivity in local ecosystems.

Hunter-gatherers are well integrated into the environment, and while their existence may seem "primitive" to us today, their intelligence is fully engaged in the tasks of knowing the landscapes, the properties and locations of plants, and the behaviors of animals. In fact, our brains and nervous systems today, unchanged from those of our hunting and gathering ancestors, were shaped by the long millennia of intimate dependence on the natural environment.

Successful hunter-gatherers in the past were probably mostly grouped in small, mobile bands, the energy limits described above shaping distinctive population patterns. Note that hunting and gathering societies were small and for the most part stable; mortality levels were relatively high and average life spans short, but fertility levels comparatively low. Mobility, the key to survival in a changing ecosystem, determined birth rates; women had to carry young children and could not feed or care for larger families. Probably both the energy drain of being on the move and the extended breast-feeding of young children acted as natural birth control by suppressing ovulation. Overall, the characteristic pattern of population dynamics would be one in which both birth rates and death rates fluctuated, but birth rates never rose high enough or remained high long enough to generate long-term population growth.

With these energy limits and population patterns, hunting and gathering bands tended to disappear in bad years or when they failed to reproduce, but they could not easily overharvest or permanently damage the ecosystem that sustained them. Exhausting a local food resource, a group would simply move on, giving the ecosystem a chance to regenerate. But when hunting and gathering groups first settled, as the Natufians apparently did in the hills of Israel, Jordan, Lebanon, and Syria, the balance between humans and environment was disrupted in subtle but critical ways.

For the first time, as "The Environmental Revolution" shows, the population balance shifted in favor of fertility as births increased and weaning time declined. But the local environment near the Natufian settlements, harvested beyond its sustainable yield, apparently began to decline; its capacity to support human beings diminished. Rising scarcity and environmental degradation inevitably prompted a change. For the Natufians, it was probably a change back to nomadism and a gradual disappearance from the archaeological record. Other groups in the Middle East responded to scarcity in a different way—by changing the relationship between humanity and the earth fundamentally for the first time, with the invention of agriculture.

The first agricultural societies were supported and limited by sunshine and by the ability of plants to capture and convert it into harvestable, digestible growth. But deliberate planting of edible crops changed the energy balance and made available a reliable share of plant growth for the support of human beings. Agriculture, which required field work at particular times during the year, enforced settlement; the human labor devoted to farming created something that had not previously existed, the possibility of a *surplus* that could be stored to feed the community until the next harvest. Food storage was a modest but crucial expansion of the energy constraints on human society. Settlement and surplus together caused an equally revolutionary change in population dynamics.

Settled farming communities like the one that existed at Ain Ghazal achieved a new ratio of births and deaths, one in which birth rates consistently exceeded death rates, possibly for the first time in human experience. The rate of expansion was not rapid, certainly not by present standards. At the end of Ain Ghazal's history, about 6,000 years ago, the population was apparently doubling every 500 years, an annual growth rate of 0.14%, a rate that today would be considered near zero population growth. (For comparative purposes, the U.S. population today is growing by 0.7%, a rate sufficient to

double the population in a century.) But sustained in a settled community over centuries, even a slow rate of increase changes the relationship between humans and resources.

In Ain Ghazal's case, the population–resource imbalance led to disappearance of the community, possibly due to the inevitable deterioration of the land under cultivation—soil exhaustion. The point of the example is that once again growth inevitably results in scarcity of food and energy, which prompts change. Change may simply mean disappearance, or it may mean a shift to a completely new way of life that once again redefines the energy limits and population dynamics of human society. In the ancient Middle East, there was no simple way to significantly expand the land's productivity, the amount of food harvested from each acre. No such change would occur for thousands of years. The story of agriculture and civilization is mostly one of expanding farming to new lands and adapting society to those conditions.

Clearly the Industrial Revolution was a change without precedent in human experience. Coal, the first fossil fuel in widespread use, vastly expanded the energy limits on human society. Unlike the wood used for heating and cooking or the crops harvested as foodstuffs, coal was not a direct product of the natural ecosystem. The fossilized remains of plants and animals accumulated and concentrated over millions of years, coal was a sort of "bank account" of the natural productivity of earlier eras, suddenly available for present use. It was available in a concentrated form that could be used in ways never previously imagined. In addition, coal could be used in ways that substituted for human and animal muscle power—in the factories and in new farming equipment. It was like a sudden increase in the pool of humans and animals available for labor, an increase that needed neither pasturage nor additional food supplies to sustain it.

The key to understanding the population dynamics characteristic of industrial societies is the observation that *living standards matter*. As the examination of marriage records in the English village of Colyton shows, fertility levels in England actually *increased* as the Industrial Revolution took hold there and families responded to new sources of material wealth. England's century of population growth was caused by a shifting balance between birth rates and death rates, as always, but not (at first) because death rates slipped below birth rates.

Instead, fertility rates climbed and stayed high. When conditions worsened, marriages declined and fertility levels fell. In this case, a human population clearly assumed some measure of control over its demographic destiny, and that control was responsive to living standards.

The Industrial Revolution changed humanity's relation with the earth more fundamentally than any previous change in human history. It changed the amount of energy and resources available to support each human life, and it smashed (temporarily) the limit imposed by the annual flow of energy from the sun. You should note, as you compare this recent revolution with the earlier invention of agriculture, that the Industrial Revolution also came about partly as a response to scarcity caused by overuse of the environment—in England's case, by the overuse and disappearance of wood fuel. Wood was needed for heat, for shipbuilding, and for the charcoal used in prevailing small-scale manufacturing. The substitute was surface outcrops of coal in the northern landscape—smoky, dirty, but available. Eventually the surface supplies of coal were exhausted, forcing miners to dig deeper—and leading to the application of coal-fired steam engines to pump water from the mines, the innovation that turned the corner of the industrial era.

As you compare hunter-gatherers, settled farmers, and industrial areas in terms of energy limits, population dynamics, and the response of each type of society to scarcity, remember that the three levels do not represent an inevitable progression or advance from "primitive" to "modern." Hunting and gathering, agriculture, and industrialization represent three very different ways of exploiting the earth's resources. All three exist today, although in different proportion than any time in the past. There are hunter-gatherers in Africa, Asia, and Latin America today, although such native people are often under great pressure from outsiders seeking the resources on their lands. And hundreds of millions of people around the world, a majority in many countries, support themselves with subsistence farming practices not so different from the methods of the first farming communities in the Middle East. In the case studies shown in the units ahead, ask yourself what level of human impact is being shown and how each level becomes unstable or unsustainable.

The human species today, in the midst of an unprecedented population expansion, is by far the most numerous large mammal ever to live on earth. The gradual tide of population growth that began when the first societies settled and started farming has now risen in an extraordinary wave. Virtually all of that wave is accounted for by growth in agricultural and industrial societies. As we consider the impacts of earth's present population of 6 billion and the growth in prospect, think back to the examples shown in "The Environmental Revolution" and consider how growth has inevitably undermined the pattern of resource use characteristic of each type of society. How the world will accommodate the growth now taking place is one of the most important questions ever to face the human species.

The Industrial Revolution: A Closer Look

The examples of England's Industrial Revolution shown in the television program illustrate two themes that we will encounter repeatedly throughout *Race to Save the Planet*. One is the reinforcing interplay of factors responsible for accelerating changes, and the second is the environmental impacts associated with the technologies and patterns of resource use characteristic of the industrial era.

Notice how the examples from England reveal the tight interdependence of energy sources and technologies. Discovering or developing a new energy source inevitably creates an impetus for use of the technologies designed to harness that source, often with unexpected results. In England's case, the coal-powered steam engine permitted water pumping from underground mines, making coal cheap and plentiful and guaranteeing that the steam engine itself would be put to a range of previously unimaginable tasks. Coal stimulated a new technology (the steam engine), and the technology was first applied to increase the availability of coal. Just as important were the new iron-smelting technologies that coal made possible. In turn, the new, cheap cast iron found application in a host of new industrial settings. This self-reinforcing dynamic has been repeated in many forms throughout the industrial era.

You should also note the interplay and interdependence of advances in industry and agriculture that took place in England. The new way of harnessing energy and transforming raw materials using fossil fuels might not have been enough to cause a complete shift in England's economy to industrial manufacturing, without the changes in farming and livestock-raising practices that had taken place a little earlier. New planting practices such as the Norfolk four-course rotation and selective breeding of livestock allowed farmers to use their animals and land more effectively and to produce bigger harvests with fewer men. The change in England's agricultural economy freed farm workers to take jobs in the new factories, reinforcing the spread of industry and the establishment of factory-based manufacturing. Industrial and agricultural change typically go hand-in-hand.

Finally, changes in land and resource use also change settlement patterns, in ways that can be mutually reinforcing. The use of coal as a power source meant that manufacturing plants could be clustered in cities rather than scattered on the landscape to capture more dispersed power sources such as flowing water; the concentration of plants meant a new concentration of employment, encouraging the rapid growth of cities; and this new urbanization brought with it new environmental and health problems. As countries like England became more urbanized, use of fossil fuels became even more essential to support the urban masses, problems of waste disposal grew more severe, and the dynamic of the Industrial Revolution took another turn. In later units, think about the relationship between settlement patterns and environmental problems— how the kind of society we have determines the kinds of problems we confront and shapes the solutions we can consider.

Even in the early years of the Industrial Revolution in England, certain environmental impacts that would become the hallmark of the industrial era had become apparent. The first of these was damage to ecosystems, as revealed by the studies of sphagnum moss in the northern moors shown in "The Environmental Revolution." The coal smoke from factories in cities such as Sheffield and Manchester, settling across the moors, gradually poisoned native mosses. The ecosystem itself thus became a living record of the industrial age, its characteristic plants showing the toll of early industrial pollution. We will encounter several other examples of such "bioindicators" of environmental damage in later units.

Second, the link between environmental quality and human health first became clear during the early days of the Industrial Revolution—first as the health toll due to contagious disease, encouraged by urban overcrowding and inadequate waste disposal in industrial cities. In Units 4 and 5, we will see how deterioration of air and water quality inevitably affects human communities and how the deterioration has become global in scope. In Unit 10, we will look more closely at the problem of waste and new methods of waste reduction and management that, among other things, controlled the risk of contagion that has always plagued urban settlements.

Attitudes Toward the Earth

"The Environmental Revolution" documents a third transition rather different from the agricultural and industrial revolutions—a far-reaching contemporary change in attitudes about environmental damage. In the early industrial era, obvious evidence of environmental damage was met with a faith in technology's ability to repair the damage. The fouling of the Thames River with sewage, shown in the program, eventually led to a highly engineered drainage system for waste disposal, a project that demonstrated both the spirit and sophistication of British technology. The Thames example is symbolic of an attitude that prevailed for more than a century, that human damage to land, air, and water could be reversed by technology, that problems would inevitably call forth solutions from the ingenuity and innovation of the times.

In the early 1960s, an attitude without precedent in industrial society began to take hold in the United States: an environmental awakening in which the faith in technology began to erode. Evidence of global scale environmental contamination—the strontium-90 fallout from early nuclear bomb tests and the measurable traces of the pesticide DDT detected in the tissue of birds and animals—spread the awakening from a small circle of concerned scientists to the public at large. But like earlier transitions in human history, there seems to be no going back to the old way of looking at things.

The questioning of progress that is the central theme of the environmental awakening was compounded from several elements. First was the perception of technology itself as part of the problem, the technologies of the industrial age having become intrinsically damaging to the environment. Evidence of environmental damage and pollution shifted people's attitudes toward technology from a basically beneficial force requiring some fine-tuning to an inherently destructive agent that must be monitored and managed to minimize damage.

A second dimension of the environmental awakening is the role played by technology itself in revealing the extent of environmental damage. New monitoring technologies, instruments capable of detecting chemical contamination in low concentrations, and the satellites used to photograph the earth from space all extended the human ability to perceive the changes that industrial society had set in motion and to recognize the urgency of response.

In the preceding section, we discussed the idea that scarcity has provoked the major transitions in human impact on earth. The environmental awakening, and what in this course we call the "environmental revolution," has been prompted by a new sort of scarcity—scarcity of resource quality. Environmental awareness and action have been driven by evidence that the earth's land, waters, and atmosphere have a limited capacity to assimilate the wastes and materials of the industrial age. *Race to Save the Planet* is about how humanity is adjusting to this new awareness of scarcity—the scarcity and diminution of the earth's productive capacity by a population of unprecedented size and the scarcity of the earth's capacity to assimilate the effects of an industrial technology based on fossil fuels.

The environmental awakening that began less than three decades ago—the *awareness* of global environmental change—is only the first step in a far-reaching readjustment of human activities. The remaining ten units of this course examine contemporary environmental problems and solutions as a way of sketching the outlines of the adjustment that lies ahead, the societal response to that awareness.

Examine Your Views and Values

1. Archaeologists believe that the Natufians, once they abandoned their nomadic ways and settled in permanent sites, gradually and inevitably exhausted the wild plant and animal resources they depended on. Can you think of other changes that might have made their way of life unsustainable once they settled? Consider the evidence shown in "The Environmental Revolution" that led archaeologists to their conclusion. What sorts of evidence would be needed to support a different theory of the disappearance of this culture?

2. Dr. E. A. Wrigley's study of the town records of Colyton, England, shows that marriage and childbearing patterns during the Industrial Revolution in England responded to economic conditions. How do your expectations about the future influence the number and timing of children you would like? How much influence do economic concerns have on your plans? What other considerations do you think affect people's childbearing plans in our society?

TEST YOUR COMPREHENSION

Self-Test Questions
(Answers at end of Study Guide)

Multiple Choice

1. Hunting and gathering societies are not likely to experience rapid population growth because
   a. they have well-developed artificial means of birth control.
   b. the mobility required for survival keeps fertility rates low.
   c. an unhealthy diet keeps fertility rates low.
   d. they know that stable populations mean balance with the environment.
   e. their customs delay marriage until late in life.

2. Hunting and gathering societies were quickly outnumbered by farming societies after the agricultural revolution because
   a. cultivated crops provided better nutrition than did wild foods.
   b. hunter-gatherers degraded their environment and disappeared.
   c. fertility levels rose in settled communities and populations began to increase.
   d. the death toll from disease was eliminated in agricultural communities.
   e. farming societies made better use of the natural environment.

3. The Natufian sites shown in "The Environmental Revolution" provide evidence that
   a. settled hunting and gathering societies could overuse their environment.
   b. hunting and gathering cultures were inevitably destined to fail.
   c. these people were practicing farming about 12,000 years ago.
   d. a settled community could not be sustained without fossil fuels.
   e. nutritional levels improved as people gave up hunting and gathering.

4. The excavation of villages at Ain Ghazal suggests that societies first made the transition to agriculture because
   a. they overcut forests and needed a better use of the cleared land.
   b. people preferred work on farms to the difficult job of tracking game.
   c. their cities needed surplus food that only farming could supply.
   d. population growth meant that for the first time they had enough labor.
   e. they gradually exhausted supplies of game and wild plants.

5. The rates of population growth in early agricultural societies were
   a. faster than any growth rates today.
   b. about as fast as the rates in developing countries today.
   c. about as fast as populations we would consider "stable" today.
   d. slower than the earlier growth in hunting and gathering societies.
   e. about as fast as the growth in England during the Industrial Revolution.

6. The first critical use of fossil-fuel–powered machinery in the Industrial Revolution was
   a. construction of the iron bridge over the Severn Gorge.
   b. steel smelters using new high-quality coal.
   c. railroads using the steam-powered locomotive.
   d. draining underground coal mines using steam-powered pumps.
   e. mass production of textiles using steam-powered weaving looms.

7. Research on the pattern of population growth in England during the Industrial Revolution shows that
   a. population expanded because death rates fell dramatically.
   b. marriage rates and fertility rates responded to living standards.
   c. population growth undermined the progress of industrialization.
   d. food production couldn't be expanded as rapidly as industrial output.
   e. population growth was rapid because more children meant more workers.

8. The United Nations Conference on the Human Environment, held in Stockholm in 1972, was significant because
   a. developing countries and industrial countries were largely in agreement about the most important environmental problems.
   b. for the first time since World War II, nations set aside their political differences to achieve a common objective.
   c. environmental activists who organized Earth Day in 1970 ran the session.
   d. it provided the first global forum for dialogue on environmental problems.
   e. the assembled delegates voted in favor of controlling population growth.

True or False

1. England's shift to reliance on coal was prompted by the exhaustion of wood supplies. _____

2. When the Industrial Revolution began in England, the country had not yet sustained any environmental damage. _____

3. Pollution by the new industries in England was welcomed by many people as a sign of prosperity. _____

4. The first evidence of pollution in England was declining game populations. _____

5. Workers joined the new industries because farming was dying out. _____

6. Changes in agriculture were as important as changes in manufacturing methods in the Industrial Revolution. _____

7. The two key developments that encouraged the environmental awakening of the 1960s were fallout of strontium-90 from bomb testing and publication of *Silent Spring*. _____

8. Although Rachel Carson's book was influential, her claims about pesticides were proved to be exaggerated. _____

9. Once people recognized the dangers of DDT, the pesticide was quickly banned. _____

10. The Yannacones' lawsuit to ban spraying of DDT on Long Island was the first significant use of the courts for environmental protection. _____

11. Earth Day 1970, though a popular success, was ignored by most politicians. _____

12. The environmental movement in the United States was based on progress against pollution in Europe. _____

Sample Essay Questions

1. Why do anthropologists believe that hunting and gathering people first turned to agriculture?

2. How do the three types of human societies (hunting and gathering, agricultural, and industrial) differ in the types and quantities of energy used?

3. Explain how populations can continue to grow even when total fertility rates fall below replacement level.

4. Use the relationship between birth rates and death rates to explain how population growth in developing countries in the late twentieth century differs from population growth in England during and after the Industrial Revolution.

5. Contrast the attitudes expressed by industrial countries and developing countries at the United Nations Conference on the Human Environment in 1972. Why were developing countries then willing to welcome pollution?

GET INVOLVED

References

Bronowski, Jacob. *The Ascent of Man.* Boston: Little, Brown & Co., 1974.

Brown, Lester R., and Hal Kane. *Full House: Reassessing the Earth's Population.* New York: W.W. Norton, 1994.

Carson, Rachel. *Silent Spring.* Boston: Houghton Mifflin Co., 1962.

Ehrlich, Paul, and Anne Ehrlich. *The Population Explosion.* New York: Doubleday, 1990.

Gottlieb, Robert. *Forcing the Spring: Transformation of the Environmental Movement.* Washington, DC: Island Press, 1993.

Keyfitz, Nathan. "The Growing Human Population." *Scientific American*, Sept. 1989.

Milbrath, Lester R. *Learning to Think Environmentally While There Is Still Time.* Albany: State University of New York Press, 1995.

Pointing, Clive. *A Green History of the World: The Environment and the Collapse of a Great Civilization.* New York: St. Martin's, 1992.

Quinn, Daniel. *Ishmael.* New York: Bantam/Turner, 1992.

Ward, Barbara, and René Dubos. *Only One Earth.* New York: W. W. Norton & Co., 1972.

Organizations

The growth of the human population, introduced in this unit, is one of the most important challenges that humanity faces today. At the present growth rate of 1.67% per year, the human population will reach 11 billion by the year 2030, well within the lifetime of most students taking this course. The fastest-growing populations today will double in size before the year 2010.

Contact the following organizations to learn more about the study of population; the economic, social, and environmental consequences of rapid growth; and efforts underway to stabilize populations.

Population Institute
107 2nd St. NE, Suite 210
Washington, DC 20002
Tel: (202) 544–3300
web: www.populationinstitute.org

Population Reference Bureau
1875 Connecticut Ave. NW
Suite 520
Washington, DC 20009-5728
Tel: (202) 483-1100
web: www.igc.apc.org

Zero Population Growth
1400 16th St. NW, Suite 320
Washington, DC 20036
Tel: (202) 332–2200
web: www.zpg.org

Only One Atmosphere

The changes human activities are causing in the atmosphere, which may lead to unprecedented global warming within the next generation, have alerted people around the world to the urgency of environmental concerns.

BEFORE YOU VIEW THE TELEVISION PROGRAM

Learning Objectives

After completing the assigned readings and viewing "Only One Atmosphere," you should be able to

- explain the link between the human use of chlorofluorocarbons (CFCs) and the ozone hole, and discuss the risk to human beings from ozone depletion.

- describe the mechanism of the greenhouse effect and list the major causes responsible for it.

- describe the most likely consequences of rapid climate change and list several steps that could be taken to slow the rate of change.

- discuss why international agreements will be needed to respond effectively to the risk of climate change.

- explain how your own activities contribute to the causes of the greenhouse effect and list the ways in which you can be part of the solution.

Reading Assignment

Choose the material from either textbook as your reading assignment. Your instructor might assign additional readings as well, and you might also find the references listed at the end of this Study Guide unit useful.

Living in the Environment

Chapter 4, "Ecology, Ecosystems, and Food Webs" (also assigned in Unit 2)

Chapter 5, "Nutrient Cycles and Soils," Section 5-1 through 5-6 (also assigned in Unit 2)

Chapter 6, "Evolution and Biodiversity"

Chapter 9 "Community Processes," Sections 9-5 and 9-6

Chapter 10, "Population Dynamics"

Chapter 18, "Air Pollution," Section 18-1

Chapter 19, "Global Warming and Ozone Loss"

Environmental Science

Chapter 4, "Ecosystems and How They Work" (Review from Unit 2)

Chapter 5, "Evolution, Biodiversity, and Community Processes," Sections 5-3 through 5-7

Chapter 7, "Population Dynamics, Carrying Capacity, and Conservation Biology"

Chapter 10, "Global Warming and Ozone Loss"

Unit Overview

This unit is about global change: how human activities cause it and what must be done in response. We will look at changes in the earth's atmosphere and consider the implication of those changes for society today and in the future. The evidence of wholesale changes in the composition of the atmosphere, and the likelihood of an unprecedented global warming within the next generation, has alerted people to the urgent need to address environmental concerns.

The assigned readings introduce the composition of the earth's atmosphere, the major types of air pollutants, and the effects of air pollution on the stratospheric ozone layer and on global climate. The readings emphasize the role of the sun as the primary energy source for life on earth and the power source for the circulation of the atmosphere and oceans. Pay particular attention to the discussion of the global carbon cycle and the hydrological cycle, and understand the importance of these cycles to human activities and their vulnerability to human impacts. The text reviews the processes responsible for ozone depletion and global warming and outlines the steps needed to respond to these threats to planetary stability.

The accompanying television program, "Only One Atmosphere," examines the evidence that humans have caused fundamental changes in the earth's atmosphere and looks at what scientists currently know about the consequences of these changes. The program begins in Australia, where heightened rates of fatal skin cancer are coupled with concerns about damage to the ozone layer, in particular because of the Antarctic ozone thinning nearby. The program then explores the more complex problem of the greenhouse effect, investigating the basic mechanism, the tools scientists use to study it, the probable consequences of a rapid rise in global temperature, and the unresolved questions that researchers are trying to answer. The program concludes by considering the various steps that can be taken to stem greenhouse gas emissions and slow the rate of global warming.

The changes unfolding in the earth's atmosphere provide powerful evidence that human impact on the environment has reached a new stage. For the first time in human history, the problems have become unmistakably global and cannot be addressed successfully by any one nation acting alone. Atmospheric change is symbolic of other environmental changes that are shaping a new awareness of the common threat of environmental degradation and the urgent need to develop new, cooperative approaches to maintain the earth's habitability. In later units we will look more closely at other global changes and the innovative responses emerging around the world as a new commitment to common survival takes shape.

Glossary of Key Terms and Concepts

The following terms and concepts will be useful as background for viewing "Only One Atmosphere."

The earth's **atmosphere** is a multilayered mixture of gases and water vapor that clings to the earth's surface in a layer about 40 miles thick, proportionally no thicker than the skin of an apple. Many of the gases are involved in chemical cycles such as the nitrogen cycle and carbon cycle that sustain life on earth and shape the planet's habitability. Nitrogen and oxygen, the most abundant gases, make up more than 99% of the atmosphere, while so-called trace gases including carbon dioxide, methane, and other "greenhouse" gases (see below) constitute the remainder.

Chlorofluorocarbons (CFCs), better known under the trade name Freon, are a class of synthetic chemicals first manufactured in the 1930s and widely used as refrigerants, spray propellants, solvents, and blowing agents for plastic foam. Stable and inert at ground level, these chemicals drift to the stratosphere where the sun's rays break apart the chlorine and fluorine atoms they contain, causing a series of reactions that results in destruction of stratospheric ozone.

Climate models are elaborate computer programs that simulate the interplay of the sun's energy with the earth's land surface, oceans, and atmosphere. By changing equations in the program that represent factors such as the mix of gases in the atmosphere or the amount of snow and ice on the earth's surface, scientists can investigate how the earth's climate (rainfall and temperature patterns) might change over time in response to human activities.

The **greenhouse effect** is the name given to the phenomenon by which the transparent gases of the atmosphere allow sunlight to warm the earth's surface and then prevent that warmth from escaping back into space. Like the inside of a greenhouse on a sunny winter day, the earth is warmer because of this effect than it would be if the atmosphere did not hold heat; the gases responsible for this heat-trapping effect constitute less than 1% of the atmosphere's volume.

Halons are synthetic chemicals related to CFCs that contain bromine rather than fluorine or chlorine. Widely used as flame retardants in fire extinguishers, they are also linked to damage to the earth's ozone layer.

Malignant melanoma is an often-fatal form of skin cancer that can be caused by exposure to high levels of ultraviolet light, for instance, in regions over which the ozone layer has been damaged. Scientists fear that this form of cancer is likely to become more common if deterioration of the ozone layer continues.

Methane is a natural gas that results from biological fermentation in the absence of oxygen, produced in the digestive tracts of livestock and termites, in flooded rice paddies and marshes, and in landfills used for trash disposal. Present in low but rapidly increasing concentration in the earth's atmosphere, methane is a potent greenhouse gas that traps heat and contributes to global warming.

Found in offshore sediments in the Arctic Ocean and buried beneath the surface of the tundra, **methane hydrate** is a stable complex of methane and ice. If ocean temperatures warm slightly and tundra thaws as predicted by the greenhouse effect, the complex may melt and release large quantities of methane into the atmosphere, accelerating global warming.

Ozone is an unstable and chemically reactive gas containing three oxygen atoms, formed at high altitudes by the action of sunlight on molecular oxygen. Present at low concentration in a layer high in the stratosphere, ozone absorbs ultraviolet radiation from the sun and reduces the amount of this damaging radiation that reaches the earth's surface. Ozone is also formed at ground level by the interaction of sunlight with exhaust gases from automobiles and industry, a problem that is discussed in Unit 5.

The **"ozone hole"** —which is actually a thinning, not a hole—is the popular name given to a phenomenon discovered in 1987, when scientists measured unexpectedly low ozone concentrations in the stratosphere above the South Pole during the Antarctic spring. The loss of ozone is now known to be caused by chemical reactions set in motion by CFCs.

The spectrum of light emitted by the sun includes **ultraviolet radiation**, which, though not visible to humans, contains enough energy to damage plant and animal cells. Exposure to high levels of ultraviolet light can cause skin cancers, blindness, and immune system problems in mammals and leaf damage and low crop yields in some types of plants.

AFTER YOU VIEW THE TELEVISION PROGRAM

Consider What You Have Seen

"Only One Atmosphere" concerns some of the most difficult scientific and public policy challenges that humanity has ever faced. As you reflect on the case studies presented in the program, try to relate them to the concepts discussed in the reading assignment. Consider the following questions.

- What do we know?

- How do we know?

- What can we do?

- Who pays, who benefits?

What Do We Know?

The science of global change is relatively young. In the case of ozone loss both the chemical reactions involved and the actual evidence of damage have been revealed only within the last 20 years, with many of the most important findings emerging since 1985. The basic mechanism of the greenhouse effect was first described nearly 100 years ago, and regular measurements of the atmosphere's rising concentration of carbon dioxide were begun in 1957, but it has only been within the last decade that scientists began to combine several different types of research to illustrate the risk of climate change.

The mechanisms involved: The phenomenon of ozone depletion is now one of the best-understood chemical processes in the atmosphere. Recall from the program that the first changes in ozone were measured over the polar regions. Unlike many chemical reactions, the process by which chlorine (from CFCs) breaks apart ozone molecules is speeded up by very cold temperatures—because the reaction takes place on high-altitude ice particles. Compare this process with the greenhouse effect, which involves not chemical reactions but rather an accumulation in the atmosphere of gas molecules that trap heat and let temperatures rise. Make a list of the chemicals believed responsible for ozone damage and for the greenhouse effect. How are these two problems linked?

Chemical cycles: The chemicals in the earth's atmosphere don't remain there indefinitely. Most are involved in cyclic exchanges among the atmosphere, oceans, and living organisms known as **biogeochemical cycles**; some of these are described in your textbook. Be sure to understand the carbon cycle and the nitrogen cycle. At one point early in "Only One Atmosphere," the narrator refers to coal, oil, and natural gas as "fossil sunlight." See if you can explain this phrase with reference to the carbon cycle.

The CFCs are synthetic chemicals that never existed in nature before they were first released by human activities. They are not part of a natural chemical cycle. They do demonstrate a phenomenon that is a concern for all types of pollutants of air and water: **residence time**. CFCs are exceptionally long-lived chemicals because they participate in few chemical reactions. Once released from spray cans or leaked from air conditioners, they remain in the area of the atmosphere closest to the earth's surface, called the **troposphere**, for up to eight years. They gradually rise to reach the upper layer, the **stratosphere**, where they may last intact for up to 100 years. It is only after long exposure to ultraviolet light at those altitudes that chlorine splits off and the chlorine atoms begin to destroy ozone molecules. This long residence time, coupled with the exponential increase in production and release of CFCs since their discovery in the 1930s, means that CFC levels would continue to rise, and ozone damage increase, even if all CFC production were halted immediately. This cumulative, long-term dimension has heightened the urgency of curtailing CFC production worldwide.

The importance of the past: "Only One Atmosphere" cites several cases in which scientists have used information about past climate changes to better understand the speed and likely impacts of climate change due to the greenhouse effect. Recall how scientists used carbon dioxide concentrations measured in air bubbles trapped in glaciers in Greenland and Antarctica to check whether current climate models could account for conditions during the last ice age. What other examples from the program seek to apply an understanding of past climate changes to the earth's present situation?

Human impact: All of the gases believed responsible for the greenhouse effect are so-called trace gases that make up only a tiny proportion of the atmosphere by volume. Review the discussion of atmospheric composition in your reading assignment. The gases mentioned in the program—carbon dioxide, methane, and CFCs—make up only a few hundredths of one percent of the atmosphere's volume. This very scarcity makes them susceptible to human impact. Another point to keep in mind is that the different gases have different "greenhouse strengths"—some are far more effective at trapping heat than are others. One hundred molecules of CFCs, for example, trap as much heat in the atmosphere as one million molecules of carbon dioxide.

The changes human activities have caused in the atmosphere are recent, all traceable to the few human generations since the Industrial Revolution.

Many of the most important changes have occurred just since the end of World War II. The concentration of carbon dioxide, for example, has risen about 30% since the Industrial Revolution, but the increase that occurred in the last 30 years exceeds that in the previous 200. CFCs were created in 1931 and their use slowly and steadily increased, accelerating sharply after World War II; today they are increasing more rapidly than any other greenhouse gas. The problems these gases pose have never before been faced.

How Do We Know?

The science of atmospheric change is rapidly maturing, and "Only One Atmosphere" shows several of the most important techniques used to investigate the earth's climate. Two essential and interrelated dimensions of scientific work are presented: the gathering of direct observations and measurements, and the creation of models of phenomena that allow researchers to test their hypotheses.

Measurements and direct observations: What did scientists have to measure in order to confirm that damage to the earth's ozone layer was occurring? Why did they choose to work in polar regions? What observations are needed to confirm the greenhouse effect? Why are most scientists reluctant to say that global warming has begun, even when they accept the greenhouse theory?

Climate models: The use of computers has revolutionized the study of climate and understanding of the greenhouse effect, and yet climate models leave many important questions unanswered. Climate models offer scientists a way to go beyond their observations and measurements and to look into the future or the past. Review the different uses of models presented in "Only One Atmosphere." What do present climate models predict about the effects of a doubled concentration of carbon dioxide? How are models used to "predict the past"? Why do such tests raise scientists' confidence in the accuracy of their models? What are some of the factors that climate models do not yet incorporate that may influence predictions about future climate change?

What Can We Do?

Acting in the face of uncertainty: It should be clear to you after viewing "Only One Atmosphere" that scientists will never be able to predict exactly when the warming of the earth will be unambiguous, when human responsibility for climate change will be clear for all to see. Yet most scientists agree that waiting too long to slow the buildup of greenhouse gases could be disastrous. A central aspect of this issue and many others we will investigate in *Race to Save the Planet* is the need to address a problem that is not yet fully understood. How does climatologist Stephen Schneider, interviewed in "Only One Atmosphere," justify his belief that society should take action now on the causes of the greenhouse effect?

Key response strategies: The program describes five key steps that the United States could take in order to reduce the gamble of climate change—steps that would slow the rate at which we release greenhouse gases and thus delay the date at which catastrophic global warming would be likely. List the five measures, and briefly jot down how each of them would affect your life. What changes would you experience as an individual? What steps can *you* take regardless of what U.S. political leaders decide to do? What would motivate you to do so?

The limits of adapting: Some people believe that while the earth's climate may be changed by human activities, societies will be able to adapt as they have to changes in the past. What is some of the evidence presented in "Only One Atmosphere" that societies' ability to adapt will be limited? Of the three major challenges shown in the program—rising sea levels, disruption of food supplies, and heightened competition over water supplies—which would be the most difficult to respond to? Why do you think so? Which one(s) might affect your life? How?

Series themes: "Only One Atmosphere" raises many issues that will be dealt with in more detail in later units of the *Race to Save the Planet*. Unit 7, "Remnants of Eden," looks more closely at how the earth's biological diversity is being affected by global changes; Unit 8, "More for Less," explores how

society's energy needs could be supplied in a way less damaging to the atmosphere and environment; Unit 9, "Save the Earth—Feed the World," considers the major challenges to world food production and how farmers are changing their methods; and Unit 12, "Now or Never," looks at how some individuals and international institutions are taking the lead in forging responses to global change. As you work on these later units, you may wish to review the fundamentals of the greenhouse effect and atmospheric change presented here.

Who Pays, Who Benefits?

The responsibility for atmospheric changes is not shared equally by all societies, and the consequences of climate change will be far more disruptive for some societies than for others. These differential causes and consequences have to do with levels of affluence, geographic location, and even the timing of change. As you review the material in "Only One Atmosphere," consider the following questions that ask you to think about issues of distribution.

- Which causes of atmospheric change are mostly due to practices in industrial societies? Which are due to practices in developing countries? For which causes do all societies share responsibility?

- Which consequences of atmospheric change are most likely to be severe in tropical countries? Which will be most severe in temperate-zone countries such as the United States? Which consequences will affect populations living closest to the North and South poles?

- Which consequences of atmospheric change will affect coastal and island dwellers more than inland populations? What kinds of precautions or protective measures will they have to consider?

- What effects of atmospheric change might we see during the 1990s? What effects will not be evident for a generation or more? How much responsibility do you think the present generation bears for the conditions in which its descendants will live?

"Only One Atmosphere" emphasizes the need for international agreements and cooperative efforts to respond to the challenge of atmospheric change. The fact that neither responsibility for the problem nor vulnerability to the disruptions it will cause are distributed equally (or even proportionally) among societies makes designing such agreements especially difficult. What responsibility do you think industrial countries, which have caused most greenhouse gas emissions so far, have to help developing countries to understand the problem and control their releases of carbon dioxide and other greenhouse gases? Why should industrial countries care what steps are taken in other parts of the world? We will have many opportunities to consider this question of common responsibility throughout *Race to Save the Planet*.

Take a Closer Look at the Featured Countries

The Commonwealth of **Australia** occupies an entire continent about the size of the United States located between the South Pacific and Indian oceans (Figure 4-1). The country's major cities lie in the south and east, where they are about as close to the South Pole as New York City is to the North Pole. The 19 million Australians, mostly descended from British settlers, live in an affluent industrial society with a material standard of living comparable to western Europe. The population is growing by 0.7% per year. The vast interior of Australia is desert, and in most of the country the climate is dry and sunny throughout the year.

Like all industrial societies, Australia is heavily dependent on fossil fuels for transportation and energy production. Though the country's population is small, fossil fuel burning makes it a significant contributor to global carbon dioxide buildup. Australia released an estimated 65 million tons of carbon in 1987, nearly three times as much as the country had produced as recently as 1960. (Most estimates of carbon dioxide emissions are based on carbon weight, calculated from the chemical composition of the fossil fuels burned. Each ton of carbon is carried into the air in 3.7 tons of carbon dioxide.) About four tons of carbon are released

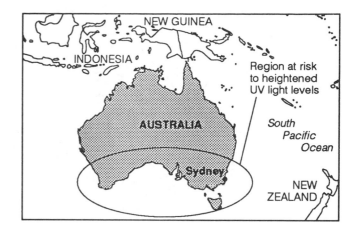

Figure 4-1   Australia

per person each year in Australia, one of the highest per capita contributions to the greenhouse effect of any society.

As shown in "Only One Atmosphere," Australia was one of the first countries to be affected by global atmospheric change (increased skin cancers traced to ozone loss), and this sense of vulnerability has perhaps heightened the country's resolve to take action on climate change. Prime Minister Robert Hawke announced a new environmental policy in the summer of 1989 that included a plan to plant 1 billion trees in Australia by the year 2000, helping to absorb excess carbon dioxide. Australia also joined other nations at Helsinki, Finland, in May 1989, pledging to phase out its production and use of the most ozone-damaging CFCs by the year 2000, and has been a strong supporter of plans to draft an international treaty on atmospheric change. Though not a major contributor to the problem, Australia can lead by example, advocating international policies and adjustments needed to control global warming.

Djibouti, a tiny desert nation sandwiched between Ethiopia and Somalia on the Gulf of Aden, lies at the same latitude as the vast Sahara Desert to the west (Figure 4-2). A desert country unsuited to agriculture, Djibouti's economy is dependent on its port on the gulf, through which most of Ethiopia's exports travel. The country's population numbers just 700,000, 80% of them living in the port city but many leading a nomadic existence dependent on small herds of goats and sheep. Djibouti's population is growing at about 2.3% per year, a rate that will double the population in about 30 years. Living

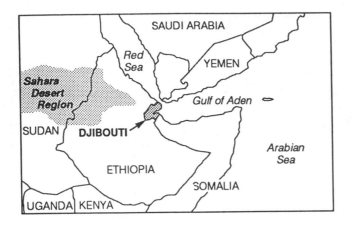

Figure 4-2   Djibouti

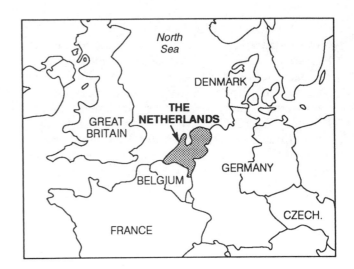

Figure 4-3   The Netherlands

standards are typical for arid nations in Africa, with a life expectancy at birth of about 48 years and 41% of the population made up of children younger than 15 years.

One of the least populous nations on earth, Djibouti is not surprisingly *not* a major contributor to atmospheric change. The country is not a heavy user of fossil fuels and has no forests whose burning could contribute to global warming. However, even Djibouti is contributing to our present understanding of climate change. The research shown in "Only One Atmosphere" has revealed that as recently as 9,000 years ago Djibouti was dotted by lakes and supported farming communities whose grain crops were supported by seasonal rains. The earth's orbit shifted, the monsoon rains disappeared, and the country's comfortable climate gradually dried out. Research in Djibouti has shown how climate can be affected by changes in the earth's path around the sun and also how societies can adapt to change that unfolds slowly, over thousands of years. Parts of the earth may face change of a similar magnitude within the next human lifetime, a span too short to make smooth adjustments.

The **Netherlands** (Holland), located in northern Europe on the North Sea, is a wealthy and progressive industrial society of 15.8 million people (Figure 4-3). Although its population is growing very slowly at 0.4% per year, the Netherlands is one of the most densely populated areas of Europe. Two-thirds of the country's land lies near sea level, and much land has been reclaimed from the sea by drainage and construction of elaborate dikes. With an income level of about $25,830 for each Dutch citizen, the country has been affluent enough to

afford sophisticated protection from storms and high tides; the proportion of the country's gross national product devoted to building and maintaining dikes actually exceeds the proportion of U.S. GNP devoted to national defense.

Like other European nations, Holland's dependence on fossil fuels and industrial technologies makes the country a significant contributor to atmospheric change. In 1987, the burning of fossil fuels in Holland released an estimated 36 million tons of carbon emissions, or 2.5 tons per person—a level somewhat lower than Europe's worst emitters of greenhouse gases, but still higher than, for example, France (which relies heavily on nuclear power) and Norway (which uses nonpolluting hydropower). In addition, Holland still has close economic and political ties to Indonesia, a former Dutch colony and now the world's second largest source of carbon dioxide from tropical deforestation.

Holland's acute vulnerability to sea level rise and its well-educated, progressive population have both served to heighten the country's sensitivity to the threat of global warming. Already a leader in recycling and one of Europe's most energy-efficient nations, Holland's government is considering proposals to invest much more in energy efficiency in order to freeze and gradually reduce the country's carbon emissions. It has supported international efforts to control CFCs and plans to phase out all use of CFCs by the year 2000, going beyond current international agreements. Holland is among the strongest advocates of negotiating an international

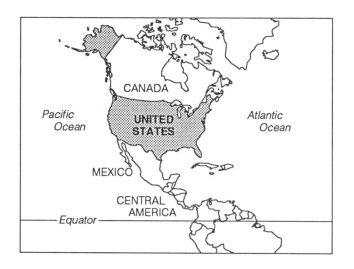

Figure 4-4 The United States

treaty on the atmosphere that would set up a common framework for stabilizing the atmosphere and coping with global warming. In both words and deeds, Holland is tackling some of the difficult adjustments industrial societies must make in the age of global change.

The **United States**, with a population of 273 million, places the world's heaviest national burden on the global atmosphere (Figure 4-4). The country's size, affluence, and dependence on coal and oil combine to account for the annual release of 1.2 billion tons of carbon, about 20% of the world's total release of the gas from fossil fuels. In addition, the country releases more carbon per person, over 5 tons in 1987, than any other society on earth. The United States is also the world's largest producer of chlorofluorocarbons, gases that damage the ozone layer and account for an estimated 40% of the U.S. contribution to the greenhouse effect. And the return of carbon-absorbing forests to many parts of the country during the twentieth century has been reversed by the continuing spread of suburbs, industrial parks, and commercial development.

The U.S. scientific community has made perhaps the greatest contribution to studying the global atmosphere and understanding the process of climate change. U.S. scientists were the first to trace the changing concentration of atmospheric carbon dioxide and to work out the chemical reactions by which CFCs break down ozone. The National Center for Atmospheric Research in Colorado is the world's foremost center for climate modeling using

supercomputers. The National Science Foundation sponsored research on the "ozone hole" in Antarctica, and NASA has investigated the possibility of ozone depletion in Arctic regions. American scientists collaborate with colleagues all over the world on understanding the dimensions of global change.

This knowledge has not yet translated into action on American responsibility for atmospheric change. The U.S. government, which neglected the problem for several years, has begun to take the risk of climate change more seriously. But progress on the steps that could reduce carbon emissions—more efficient automobiles and factories, adequate insulation in homes and buildings, comprehensive recycling of paper, glass, and metals, and development of non-fossil-fuel energy sources—is slow and uneven. U.S. chemical companies are working on substitutes for CFCs, while in other areas, powerful vested interests such as the automobile industry resist the necessary adjustments The United States supports international efforts to phase out CFCs, and members of Congress have introduced global-warming legislation that would set national goals for reduction of carbon emission, but these initiatives are still far from becoming national policy. Fortunately, public awareness is increasing quickly to the point that there may be popular support for more aggressive action on the U.S. contribution to global change.

Examine Your Views and Values

1. Climatologist Stephen Schneider compares current predictions about the magnitude and timing of global warming to a toss of a coin: Agreeing that global temperatures are likely to increase by five degrees Celsius by the year 2050 amounts to a 50% chance that warming will be greater than that and a 50% chance that warming will be more moderate. Do you agree with Schneider that a 50% chance of severe or catastrophic climate change is enough to justify taking steps now to avoid it? Why or why not? How much likelihood of an automobile accident or a household burglary does it take to convince you to buy insurance?

2. The United States, the world's most affluent society, is also the single largest cause of the atmospheric change that will affect *all* societies within the next few decades. Many other countries aspire to the freedom, mobility, and material affluence that Americans take for granted. Do you think that all societies are entitled to a standard of living comparable to that in the United States? What do you think some of the consequences would be if all people on earth lived like Americans? Do you think it is possible to say that any society has "gone too far" in its pursuit of material well-being, that some aspirations can no longer be tolerated?

TEST YOUR COMPREHENSION

Self-Test Questions
(Answers at end of Study Guide)

Multiple Choice

1. When an air mass with lowered ozone levels drifted across Australia in December 1987, the results included
   a. much higher temperatures on Australian beaches.
   b. increased risk of skin cancer.
   c. higher than usual levels of ultraviolet light from the sun.
   d. an increase in chlorofluorocarbons released by Australian consumers.
   e. More than one of the above is true.
      The correct answers are _____.

2. Malignant melanoma and other skin cancers are caused by excessive exposure to
   a. CFCs and halons.
   b. ozone.
   c. ultraviolet light from the sun.
   d. higher temperatures.

3. Over the two centuries since the Industrial Revolution, the concentration of carbon dioxide in the atmosphere has
   a. increased by 100%.
   b. stayed about constant.
   c. increased by 30%.
   d. decreased by 25%.

4. The climate models discussed in "Only One Atmosphere" look at the effects of doubling atmospheric carbon dioxide concentrations, a level that
   a. cannot actually be reached on earth.
   b. is likely to be reached by the year 2050.
   c. is likely to be reached by the year 2000.
   d. will not be reached if societies phase out CFCs.

5. The country of Djibouti was included in "Only One Atmosphere" to show that
   a. people can adapt even to extreme climate changes.
   b. global warming would be likely to bring lakes and wildlife back to the Sahara.
   c. the computer models used to predict future climate changes can also account for past climate changes.
   d. the computer models used to predict future climate changes cannot explain the lakes that once existed in the Sahara.
   e. climate changes are caused by shifts in the earth's orbit, not human activities.

6. One effect of global warming, a rise in sea levels,
   a. is now known to be unlikely.
   b. is likely because increased rainfall will run off into the oceans.
   c. is likely because water expands as it warms.
   d. may be caused by melting ice at the North and South poles.
   e. More than one of the above is true.
      The correct answers are _____.

7. Holland is included in "Only One Atmosphere" to show that
   a. although costly, it is possible to protect any country from sea level rise.
   b. sea levels have already risen in some areas because of the greenhouse effect.
   c. the costs of protection from rising seas would be beyond the reach of some nations.
   d. it will be easier to adapt than prevent the greenhouse effect.
   e. More than one of the above is true.
      The correct answers are _____.

8. If sea levels rise as predicted in Everglades National Park,
    a. the increase in wetlands would attract more rare animals.
    b. the park's animals would move to other suitable areas in Florida.
    c. towns, farms, and roads would prevent migrating animals from relocating.
    d. the area might be the first park to succumb to the greenhouse effect.
    e. More than one of the above is true. The correct answers are _____.

9. The present distribution of clouds on earth
    a. is thought to have a slight cooling effect.
    b. probably has no effect on global temperature.
    c. is thought to have a slight warming effect.
    d. is determined by carbon dioxide emissions.
    e. is something scientists have not really studied.

10. One factor that is *not* among the responses to global warming suggested in "Only One Atmosphere" is
    a. more fuel-efficient automobiles.
    b. a tax on fossil fuels to reflect the costs of warming.
    c. "scrubbers" to catch carbon dioxide in smokestacks.
    d. thicker insulation in homes and buildings.
    e. ending the use of CFCs.

True or False

1. Flooded rice paddies are an important source of carbon dioxide. _____

2. The major source of carbon dioxide is burning fossil fuels. _____

3. Deforestation causes carbon dioxide to be absorbed from the air. _____

4. Methane counteracts the effect of carbon dioxide in the atmosphere. _____

5. CFCs contribute to both global warming and ozone loss. _____

6. In 1988, the U.S. grain harvest was reduced by nearly half. _____

7. Pollination of corn plants is more efficient at high temperatures. _____

8. The greenhouse effect would compensate for diminished flow of the Colorado River by causing more rain in the U.S. Southwest. _____

9. The system to allocate water to users from the Colorado River is not designed for periods of lower flow. _____

10. Few major rivers in the world are used to capacity. _____

Sample Essay Questions

1. Explain how scientists use climate models to understand the risk of climate change.

2. Explain one major uncertainty in present models of the earth's climate.

3. Do you agree or disagree with the following statement: *The debate over global warming concerns how large it will be and how fast it will unfold, but not whether it will occur.* Provide support for your answer.

4. Ozone depletion and global warming are both complex problems that scientists are only slowly beginning to understand. Explain why the former is an easier problem to address than the latter.

5. List five responses needed to reduce the risk of climate change, and discuss their importance. What first step do you think the United States should take to address this problem?

GET INVOLVED

References

Bates, Albert K. *Climate in Crisis: The Greenhouse Effect and What We Can Do.* Summertown, TN: Book Publishing Co., 1990.

Flavin, Christopher, and Seth Dunn. *Rising Sun, Gathering Winds: Policies to Stabilize Climate and Strengthen Economies.* Washington, D.C.: Worldwatch Institute, 1997.

Gelbspan, Ross. *The Heat is On: The Climate Crisis, the Cover-up, and the Prescription.* Reading, MA: Persew, 1998.

Graedel, Thomas E., and Paul J. Crutzen. *Atmosphere, Climate, and Change.* New York: W.H. Freeman, 1991.

Makhigani, Arjun, and Kevin Gurney. *Mending the Ozone Hole: Science, Technology, and Policy.* Cambridge, MA: MIT Press, 1995.

McKibben, Bill. *The End of Nature.* New York: Random House, 1989.

Revkin, Andrew. *Global Warming: Understanding the Forecast.* New York: Abbeville Press, 1992.

Roan, Sharon L. *Ozone Crisis: The 15-Year Evolution of a Sudden Global Emergency.* New York: Wiley, 1989.

Schneider, Stephen H. "The Changing Climate." *Scientific American*, Sept. 1989.

Singer, S. Fred, and Frederick Sietz. *Hot Talk, Cold Science: Global Warming's Unfinished Debate.* Cincinnati, OH: Seven Hills.

Two chapters in *Taking Sides: Clashing Views on Controversial Environmental Issues* deal with topics covered in this unit:
.   Issue 17. Is Immediate Action Necessary to Minimize Potential Catastrophic Effects of Global Climatic Warming? (pp. 288–307)
.   Issue 18. Does the Montreal Ozone Agreement Signal a New Era of International Environmental Statesmanship? (pp. 308–325)

Also see chapters on global change in various editions of Worldwatch Institute's annual *State of the World* report:
.   Christopher Flavin and Seth Dunn, "Responding to the Threat of a Climate Change," *State of the World 1998.*
.   Christopher Flavin, "Facing Up to the Risks of Climate Change," *State of the World 1996.*
.   Christopher Flavin, "Slowing Global Warming: A Worldwide Strategy," *State of the World 1990.*
.   Jodi L. Jacobson, "Holding Back the Sea," *State of the World 1990.*
.   Cynthia Pollock Shea, "Protecting the Ozone Layer," *State of the World 1989.*
.   Sandra Postel, "Stabilizing Chemical Cycles," *State of the World 1987.*

Organizations

Most major environmental groups now have programs dealing with public education and political advocacy on global change issues. In addition, you may wish to contact one of the following groups for further information on global change research and new policy initiatives.

The Climate Institute
324 4th St. NW
Washington, DC 20002
Tel: (202) 547-0104

Environmental Defense Fund
257 Park Ave. South
New York, NY 10010
Tel: (212) 505-2100; web: www.edf.org

The Greenhouse Crisis Foundation
1130 17th Street NW, Suite 630
Washington, DC 20036
(202) 466-2823

National Center for Atmospheric Research
Office for Interdisciplinary Earth Studies
P.O. Box 3000
Boulder, CO 80307-3000
web: www.ucar.edu

Natural Resources Defense Council
1350 New York Avenue NW, Suite 300
Washington, DC 20005
Tel: (202) 783-7800; web: www.nrdc.org

Do We Really Want to Live This Way?

COMMUTER TRANSPORTATION SERVICES

By restricting industry practices and individual lifestyles—including restrictions on automobile use—the South Coast Air Quality Management District (SCAQMD) aims to reduce ozone and photochemical smog in the Los Angeles basin by the year 2007.

BEFORE YOU VIEW THE TELEVISION PROGRAM

Learning Objectives

After completing the assigned readings and viewing "Do We Really Want to Live This Way?" you should be able to

- list the causes of and describe the processes responsible for photochemical smog.

- discuss the causes and effects of atmospheric deposition of pollutants.

- use examples from the television program to illustrate how pollution control strategies have evolved from reliance on *output* control methods to *input* control methods.

- describe how risk assessment is used to evaluate pollution's hazards to human health.

- define and explain "eutrophication" of surface waters.

- compare the causes and effects of "unconventional" air pollutants in Los Angeles and "micropollutants" in the Rhine River basin.

- discuss the most current pollution control approaches being devised in Los Angeles and the Rhine River region.

- describe how your own activities contribute to problems like those shown in "Do We Really Want to Live This Way?" and explain what changes in personal behavior might reduce that contribution.

Reading Assignment

Choose the material from either textbook as your reading assignment. Your instructor might assign additional readings as well.

Living in the Environment

Chapter 17, "Risk, Toxicology, and Human Health"

Chapter 18, "Air Pollution"

Chapter 20, "Water Pollution," Sections 20-1 through 20-3, and 20-5

Chapter 26, " Sustainable Cities," Sections 26-1 and 26-2

Environmental Science

Chapter 8, "Risk, Toxicology, and Human Health"

Chapter 9, "The Human Population," Sections 9-6 through 9-9

Chapter 10, "Air and Air Pollution"

Chapter 12, "Water Resources and Water Pollution," Sections 12-5 through 12-7

Unit Overview

Pollution of air and water was the first "cause" embraced during the awakening of environmental awareness in industrial countries in the 1960s. Activists and citizens rallied around the quest for clean air and water, governments responded with landmark legislation and regulations for offending industries, and smokestack scrubbers, catalytic converters, and other "end-of-pipe" technologies became commonplace. But today, more than a quarter century later, concern about air and water pollution remains widespread, and the problem of monitoring pollutants in industrial societies and anticipating their effects has become more complex. Affluent societies, which have always relied on air and water to dilute and disperse their wastes, have learned that while these fluid media can move waste products, they cannot *remove* them. And we humans, dependent on air and water, cannot avoid the consequences of eventual exposure to the materials we release to the waters and skies.

This unit examines the pollution problems associated with affluent industrial societies—the mechanisms involved, how the problems have changed over the past 20 years, and how strategies for pollution control have slowly evolved in response. The assigned readings consider the urban character of industrialized societies, the nature of air and water pollution, and the controversial techniques used to assign risks to different levels of exposure to pollutants and weigh the costs and benefits of polluting activities. The chapters dealing specifically with air and water pollution identify the major classes of pollutants, discuss the processes responsible for pollution-related damages, and describe the methods of pollution control now in widespread use.

The accompanying television program, "Do We Really Want to Live This Way?" explores the problems of pollution in affluent societies through two case studies: the story of air pollution in the Los Angeles basin in California and the contamination of the Rhine River basin in western Europe. These two very different regions face certain common problems associated with pollution of a fluid resource—air or water—that distributes contaminants over a large area and makes direct cause-and-effect links between pollution and environmental damage or illness difficult to trace. In both places, an important shift is taking place from emphasis on capturing pollutants where they are released to curtailing the activities that generate pollutants in the first place—a shift that forces reexamination of some of the basic assumptions of industrial societies.

Key questions include: Can the unlimited mobility, consumer choice, and convenience taken for granted in industrial societies be made more compatible with the functioning of the regional and global environment? Can individuals modify their lifestyles and expectations, and societies reduce their collective impacts, to protect air and water? Can the costs of poisoned resources and damaged ecosystems be tallied accurately enough to compare them with the costs of preventing pollution?

Glossary of Key Terms and Concepts

The following terms and concepts will be useful as background for viewing "Do We Really Want to Live This Way?"

The term **acid deposition** refers to the airborne transport and descent to earth of acids and acid-forming chemicals, particularly those released by industries and vehicles. "Acid precipitation" refers to acid deposition in rain, snow, and fog.

Atmospheric deposition is the more general term used to describe any pollutants transported and delivered to earth via the atmosphere.

Bentazone is a synthetic chlorinated hydrocarbon used as a pesticide. It was detected in trace quantities in drinking water withdrawn from the Rhine River, even after extensive purification of the river water by the Amsterdam Water Works.

Fish and other freshwater organisms used as **bioindicators** are submerged in water samples taken from the Rhine River and other water supplies; their unusual behavior or death may indicate the presence of hazardous pollutants that have escaped other detection methods.

The **catalytic converter** is a device installed on the exhaust system of all new automobiles in the United States to control emissions of air pollutants, particularly those that contribute to photochemical smog such as hydrocarbons, nitrogen oxides, and carbon monoxide.

The **Clean Air Acts**, air pollution control laws passed by the U.S. Congress in 1965, 1970, 1977, and 1990 required the Environmental Protection Agency to set national air quality standards for major pollutants (primary standards to protect human health and secondary standards to preserve visibility and protect crops and buildings) and to set schedules for the attainment of air quality goals.

Eutrophication takes place in a river, lake, or other shallow body of water when polluted by runoff containing plant nutrients including nitrates and phosphates. The nutrients support an explosive growth of algae, and the subsequent decomposition of dead algae by bacteria rapidly consumes the oxygen in the water, "suffocating" fish and other organisms.

Heavy metals include copper, lead, cadmium, mercury, and other toxic metals used in industrial processes and often released as both air and water pollutants. They may accumulate to hazardous concentrations in sediments and sludge.

The **hexavalent chromium** mentioned in "Do We Really Want to Live This Way?" is a toxic heavy metal used in chrome plating processes. Exposure to airborne dust containing this metal is being investigated as a possible cause of elevated rates of cancer in parts of Los Angeles.

Hydrocarbons, air pollutants that are important precursors of smog, are chemical compounds of carbon and hydrogen generally released as unburned or incompletely burned residue when carbon-containing fossil fuels such as coal, oil, and natural gas are burned in car and truck engines.

Micropollutants are pollutants of air and water present in such tiny concentrations that they cannot ordinarily be detected by conventional means. Even small quantities of chemical contaminants such as pesticides and products formed when other chemicals break down may pose serious health risks.

The **National Ambient Air Quality Standards** (NAAQS), air pollution standards established by the U.S. Environmental Protection Agency under the Clean Air Acts, set the maximum allowable concentrations, averaged over a designated time period, for seven major pollutants in outdoor air.

Nitrogen oxides (NO_x), a class of air pollutants formed especially by burning gasoline and other transport fuels, are important precursors of both smog and acid deposition.

Nutrients include water-soluble nitrogen and phosphorus compounds (nitrates, phosphates) needed by plants for normal growth. Runoff of excess fertilizers from farmers' fields and untreated sewage are the primary sources of nutrients responsible for harmful water pollution such as eutrophication.

Ozone, an unstable form of oxygen formed near the earth's surface by the action of sunlight on nitrogen oxides and hydrocarbons, is a primary component of smog that aggravates breathing problems and damages plants. There is no connection between surface-level (tropospheric) ozone and the stratospheric ozone layer that protects the earth from ultraviolet light, as discussed in Unit 4.

Pesticides include both chemical and biological substances designed to kill plants, animals, or insects that damage or interfere with the growth of crops, timber trees, and other desired vegetation. Many chemical pesticides, transported in runoff from farmers' fields, have become serious micropollutants of drinking water supplies.

Polychlorinated biphenyls (PCBs) are synthetic chlorine-containing hydrocarbons produced as a stable insulating fluid for use in electrical transformers. They have proven to be a serious water pollutant that concentrates in the fat of living organisms; chronic exposure is linked to a variety of health effects including liver and kidney damage, miscarriages, and tumors. The manufacture of PCBs was banned in the United States in 1976, but supplies produced before then continue to contaminate the environment.

The **Rhine Action Plan**, a pollution cleanup plan adopted by the European nations that share the Rhine River (Switzerland, Germany, France, and the Netherlands), is intended to make the Rhine fishable, swimmable, and drinkable along its entire length by the next century.

Right-to-know laws require local businesses and industries to disclose information about toxic or hazardous materials used, stored, emitted from smokestacks, or disposed in the community when requested by local citizens or authorities.

Photochemical **smog** refers to a typically brown haze of pollutants in the atmosphere, formed in the presence of sunlight by reactions among primary air pollutants including hydrocarbons and nitrogen oxides released by industries, cars, and trucks.

The **South Coast Air Quality Management District** (SCAQMD) is an administrative authority responsible for air quality in the Los Angeles basin that approved a comprehensive plan to reduce ozone and photochemical smog in the region by the year 2007. SCAQMD's plan will require far-reaching changes in industry practices and individual lifestyles.

Long-term exposure to low levels of air or water pollution may cause **subclinical disease** in humans, involving a steady deterioration of body tissues and processes that does not present recognizable clinical signs and symptoms. Such exposure may accelerate the normal deterioration associated with aging or increase susceptibility to various diseases.

AFTER YOU VIEW THE TELEVISION PROGRAM

Consider What You Have Seen

The two very different regions profiled in "Do We Really Want to Live This Way?" reveal an underlying lesson about pollution into fluid media like air and water: Pollutants released anywhere in a common airshed (the Los Angeles basin) or watershed (the Rhine River basin) ultimately affect the entire basin and all of its inhabitants. As you review the stories presented in the program, consider similarities between the experience of the Rhine region and that of the Los Angeles basin. Four themes will help you see the parallels in the two case studies:

- The nature of pollutants
- Approaches to pollution control
- Evaluating risks
- Growth and the "technical fix"

The Nature of Pollutants

Your reading assignment introduces seven major classes of air pollutants and eight major types of water pollutants. The examples presented in the television program fall within these categories. The air pollutants discussed in the Los Angeles segments include

- carbon oxides
- nitrogen oxides (NO_x)
- sulfur oxides
- volatile organic compounds (hydrocarbons)
- suspended particulate matter (hexavalent chromium)
- photochemical oxidants

The water pollutants discussed in the Rhine River segments include

- oxygen-demanding wastes
- water-soluble inorganic chemicals
- plant nutrients (phosphates and nitrates)
- organic chemicals (bentazone and other pesticides, PCBs)
- sediment

You may wish to review these categories in your text to make sure you understand the distinctions between them.

The sources of pollution in the two cases have important similarities and differences. In the case of air pollutants, scientists generally distinguish two categories of polluters: **stationary sources** such as factories, power plants, and other industrial or commercial facilities with smokestacks and **mobile sources** such as cars, trucks, and buses. About two-thirds of the air pollutants responsible for smog in Los Angeles are believed due to mobile sources, while one-third are attributable to stationary sources. Sources of water pollution, in contrast, are classified as either **point sources** such as riverbank factories with identifiable discharge pipes, or **nonpoint sources** such as the runoff of plant nutrients and pesticides from thousands of acres of farmers' fields.

Strategies to control both air and water pollution have evolved over the past 25 years from approaches designed primarily to control emissions from stationary and point sources, to approaches that place considerably more emphasis on regulating mobile sources of air pollution and reducing nonpoint sources of water pollution. Mobile sources and nonpoint sources of pollutants are more difficult to monitor and more difficult to regulate than are point or stationary sources—as the television program makes clear. Contemporary pollution control efforts in, for example, southern California and the Rhine River basin involve both scientific questions and regulatory efforts of unprecedented complexity.

Another important dimension of the pollution phenomena explored in this program is the distinction between primary and secondary pollutants. **Primary pollutants** include substances harmful as they are released to the environment, such as the PCBs toxic to Waddensea seals or the airborne hexavalent chromium believed linked to cancer in parts of Los Angeles. Many pollutants of both air and water, however, are altered after their release into even more damaging forms. The most harmful components of photochemical smog, for example, are **secondary pollutants** such as ozone, peroxyacyl nitrates (PANs), and other compounds formed when nitrogen oxides and hydrocarbons react in the presence of sunlight. Oxides of nitrogen and sulfur are converted into the acid forms responsible for acid deposition after their release into the atmosphere. While primary water pollutants may also react with one another to form new damaging compounds, the phenomenon of **eutrophication** in lakes and rivers is comparable in that the damage is caused not by the nitrates and phosphates themselves, but by the biological activity (excessive growth of algae) that they stimulate.

What you should realize from this discussion of primary pollutants and secondary impacts is that release into air or water is only the *beginning* for many pollutants. Not only are air and water pollutants likely to be transported by winds or currents far from the release point, but they also may be transformed, in complex and unpredictable ways, into new configurations that alter and even amplify their ultimate damage.

Approaches to Pollution Control

As you will recall from the reading assignment in Unit 1, there are only two approaches to pollution control: **output controls**, involving methods to capture pollutants at their release point and "scrub" them from smokestacks or trap them in settling ponds, and **input controls**, which prevent pollu-

tants from entering the environment in the first place, often by restricting pollution-generating activities. Both approaches are evident in the case studies presented in "Do We Really Want to Live This Way?"

Conventional responses to pollution problems have almost always first involved output controls, "end-of-pipe" approaches that use technology to capture noxious compounds. Compliance with the National Ambient Air Quality Standards (NAAQS) set by the U.S. Environmental Protection Agency was initially seen as a matter of installing pollution control equipment on industries and automobiles; pollution control was considered a technical problem. The story of automobile emissions controls in California sums up the achievements and pitfalls of this "end-of-pipe" approach.

The first emissions control devices, required on automobiles in California in the mid-1960s, helped reduce hydrocarbon emissions by as much as 25%—at a time when hydrocarbons were considered the most serious component of smog. The oxidation catalyst, introduced in the mid-1970s, reduced hydrocarbons and carbon monoxide by a further 75% to 90%. But levels of nitrogen oxides responsible for the distinctive brown color of smog continued to grow as the number of cars and industrial sources in Los Angeles increased.

Ultimately, the solution to the problem of nitrogen oxides was another piece of technology— the three-way catalytic converter, required in the late 1970s. This time, the new technology halted the deterioration of Los Angeles air quality for nearly a decade. But this technical feat was eventually undermined by growth; the increase in automobiles in the Los Angeles basin had by 1988 offset the contribution to cleaner air by increasing the total volume of emissions. This story holds two lessons: Technical solutions to pollution sometimes involve trade offs, in which reductions of one pollutant may be offset by increased releases of another, and any technical solution may ultimately be undermined by growth.

The contemporary trend in pollution control, demonstrated by several examples in the television program, emphasizes input controls rather than further technical fixes. California's South Coast Air Quality Management District has gone perhaps further than any other authority in making input controls the centerpiece of an air pollution control plan. The SCAQMD plan involves such things as restrictions on automobile use, limits on drive-through services such as fast-food restaurants and banks, and restrictions on barbecues and other practices that people associate with the southern California lifestyle. Possibly the most controversial aspect of the plan is the fact that it affects not just businesses or manufacturers, but everyone—it forces people to examine and modify personal behavior for the societal goal of cleaner air.

Consider the parallels in the Rhine basin: Authorities in this region find themselves contending with the same sort of problem as they wrestle with nonpoint sources of water pollution. The blooms of algae in the North Sea, the first step in the process of eutrophication, are caused by many uncontrolled sources of plant nutrients including runoff from farms and vineyards, detergents, and fertilizers used on lawns and gardens. There is no simple solution to the control of nonpoint source pollutants, and ultimately control may only come through a comprehensive effort to regulate the various practices that contribute the pollutants—an effort that will touch people's lives throughout the Rhine basin.

Evaluating Risks

A major theme of "Do We Really Want to Live This Way?" involves the risks associated with various pollutants, how those risks are identified and weighed against the benefits made possible by a polluting practice, and how the risks themselves have changed as attention has shifted to new categories of pollutants overlooked by conventional control strategies.

One important thread of the story is the new attention by physicians and researchers to manifestations of health damage that could not be conclusively linked to pollution exposure using conventional clinical or epidemiological methods. (Epidemiology involves studying the pattern of disease within a community.) There are new concerns in Los Angeles and elsewhere over "subclinical" disease—the possibility that the deterioration of tissues and metabolism that we all experience as we age may be speeded up by chronic exposure to pollutants such as poor air quality. Doctors have barely begun to determine how to identify subclinical disease, let alone to understand what part specific pollutants play in causing it.

Another new and puzzling twist in under-standing the human impact of exposure to pollution is correctly interpreting pollution damage limited to small or localized groups within the general population. Epidemiologists, medical detectives, need large samples of affected groups to establish cause-and-effect links between pollutant exposure and specific health damage. But when damage appears in a neighborhood or in the immediate vicinity of a polluting facility, epidemiologists are often hard-pressed to say that apparent effects such as increased rates of miscarriage or particular cancers are clearly due to exposure to a particular chemical. When not enough people are involved to constitute a statistically valid sample, scientists and public health authorities have only guesswork to guide them. Educating people about the health risks they face becomes more a matter of speculation than certainty.

The television program shows how in some cases pollution damage to other species can be used as a warning sign of risks to humans—how these low-technology "bioindicators" may be more revealing than high-tech laboratory instruments. The accumulation of PCBs in Rhine River sediments, though measurable, was not taken seriously as a health risk until the effects on the Waddensea seals in the North Sea became widely known in 1988. In Los Angeles, the forests of the San Bernardino Mountains show unmistakable signs of damage—yellowed leaves and dying trees due to the effects of smog and acid deposition. These unambiguous ecological damages reveal how the entire system—the Rhine River basin, the Los Angeles airshed—has been changed by the pollutants poured into it.

Scientists are trying to use such bioindicators more systematically, for example, with the experiments in which fish are exposed to different types of river sediment at the Institute for Nature Management on Texel Island in the North Sea. Such biological methods are becoming more important as scientists try to detect the possible harm from new classes of pollutants that have proliferated far too rapidly to be tested adequately with conventional means. Micropollutants, such as the pesticide bentazone now being detected in the Rhine, and the bewildering array of airborne toxic chemicals in Los Angeles represent this new category of pollution risk. The chemicals in question are frequently new synthetic compounds whose possible health consequences cannot be predicted, because they are present in the environment in quantities or concentrations that defy conventional analysis or epidemiology. Watching the ecosystem closely for unsuspected species changes or damages offers the only early-warning system, albeit imperfect, that scientists have to rely on.

Growth and the "Technical Fix"

Perhaps the most important theme in this unit is that technology can offer no permanent solution to the damages caused by overdevelopment. The affluence of industrial societies can pay for innovative pollution control devices and costly cleanup, but the activities on which that affluence is based are the fundamental cause of pollution problems.

Population growth, economic growth, and consumption growth can all ultimately undermine efforts to control pollution that are based exclusively or primarily on technology. In Los Angeles, automobiles have been made far less polluting over the last two decades, but the growth of the automobile fleet associated with population growth and the region's high-mobility lifestyle have cancelled the gains bought with catalytic converters and other pollution control devices. In the Rhine valley, populations are nearly stable but the accumulated consequences of decades of economic and industrial growth are now posing risks—to food and drinking water supplies—that communities and citizens are reluctant to bear. In some cases, the problems are brought into sharper focus by bioaccumulation, or the concentration of toxic chemicals in the fat of fish, seals, and other living organisms at the top of the food chain to levels far above those found at large in the environment. Technology is important for both monitoring and controlling air and water pollution, but cannot offer a lasting solution.

In many parts of the industrial world, where the environmental and human costs of overdevelopment are coming into focus, attention is shifting to more fundamental questions that have to do with how basic human activities can be restructured to bring them more in line with the patterns and processes of the environment. The shift to pollution prevention strategies that rely on input controls, discussed in this unit, is a key part of the response. We will explore other answers in Part II of this television course, "Making Choices for the Future."

Figure 5-1   *The Los Angeles Basin*

Take a Closer Look at the Featured Regions

The **Los Angeles basin** is a sprawling urbanized region in southern California (Figure 5-1). The region served by the South Coast Air Quality Management District, discussed in the television program, encompasses Orange County and the nondesert parts of Los Angeles, Riverside, and San Bernardino counties. One of the most densely settled regions of the United States, this four-county area has a total population of about 14 million, of which the city of Los Angeles itself accounts for about 3.3 million.

These 14 million human inhabitants share the basin with an estimated 23 million motor vehicles, one of the highest ratios of cars to people found anywhere in the United States. Roughly two-thirds of the surface area of Los Angeles is paved, either roadways or parking lots to accommodate the region's automobiles. Ninety percent of the workers in Los Angeles get to their jobs by car, as there are few mass-transit alternatives. The future of air quality in the basin is tied to the future of transportation.

Los Angeles, with some of the most consistently dirty air in the United States, is frequently on the EPA's list of "nonattainment" areas unable to comply with the National Ambient Air Quality Standards (NAAQS); the city's air violated the fed-

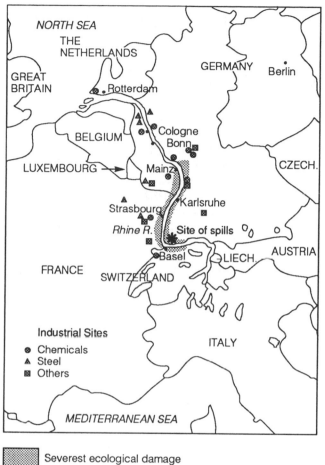

Severest ecological damage

Figure 5-2 The Rhine River Basin

eral standard for ozone on 143 days during 1985–87. It has failed to meet federal standards for four of the six principal pollutants specified by the Clean Air Acts and is the only region of the country that has not been able to meet the standard for nitrogen dioxide, produced by both auto exhaust and industrial emissions.

The new air pollution control plan designed by the South Coast Air Quality Management District is the most ambitious ever designed for an American city, and one of the most controversial. The three-stage plan is designed to bring the region into compliance with *all* federal air quality standards by 2007. Critics concede that the plan would bring cleaner air, but argue that the region can't afford an estimated annual cost of $15 billion—added expenses of about $2,500 per family—to do so. Although the air quality authority disputes the

cost, estimating that implementing the first phase will cost about $2.6 billion per year, no one argues that the economic and lifestyle adjustments required by the plan will not be substantial.

The **Rhine River watershed** drains an area in northern Europe including parts of six nations before flowing into the North Sea in Rotterdam Harbor in the Netherlands (Figure 5-2). The course of the Rhine takes it through some of the most heavily industrialized and densely settled parts of Europe, including Germany's Ruhr Valley and other areas that are the primary locations of the Swiss and German chemical industries.

The six nations drained by the Rhine have a total population of approximately 156 million; as a group, these nations are close to zero population growth. They are also among the most affluent societies on earth, with a gross national product per person of $28,250 (compared with $28,020 in the United States and $40,940 in Japan). About 40 million people live within the Rhine basin itself, including most Swiss and Dutch citizens and about a third of West Germans.

In this industrial region, the Rhine has always been a major transportation corridor and an inevitable destination for the disposal of wastes and pollutants. Yet surrounding regions also depend on the Rhine for fish and for drinking water. Long stretches of the river have been straightened for navigation; and industrial chemicals, salts from potash mines, human sewage from communities on the banks, and runoff of nutrients and pesticides from farmland have all contaminated the Rhine and its outflow in the North Sea. The river itself was considered virtually dead in the 1960s, and the North Sea now has a serious problem of eutrophication due to nutrients delivered from the Rhine.

Yet the countries that share the Rhine have always had reason to cooperate in managing the river. Since the 1800s the major countries of the watershed have signed treaties concerned with fishing and navigation rights, and in the last 25 years there have been an increasing number of intergovernmental efforts to reduce pollution of the Rhine and begin to clean it up. Some of the environmental agreements were not well enforced, but the massive spill of chemicals from a Sandoz Chemical Company warehouse in Basel, Switzerland, in 1986 galvanized public outrage over the condition of the Rhine and created political will to take stronger action to protect the river. New efforts such as the Rhine Action Plan are examples of a new commitment to address the region's long-building problems.

Examine Your Views and Values

1. One of the key themes in "Do We Really Want To Live This Way?" is that industrial societies may have reached a point in which both present levels of air and water pollution *and* new means of controlling pollution will involve lifestyle adjustments that affect all citizens. In Los Angeles and other urban areas of the United States, the only advice physicians can offer to patients with breathing problems is "Move away." Which type of lifestyle adjustment would you be more willing to accept: involuntary health impairment by pollution or mandatory behavioral changes like those included in the SCAQMD plan designed to prevent pollution? Why? Do you think governments should have the authority to impose lifestyle adjustments like those being contemplated in Los Angeles?

2. James Lents, director of the Southern California Air Quality Management District, believes that changes in individual behavior are the key to future improvements in Los Angeles air quality. Do you agree or disagree? If you agree, how consistent is *your* behavior with the kinds of changes Lents advocates in southern California? If you disagree, what evidence would you offer to Lents in support of your position?

TEST YOUR COMPREHENSION

Self-Test Questions
(Answers at end of Study Guide)

Multiple Choice

1. The photochemical smog in the Los Angeles basin is formed by
   a. chemicals released from trees in the San Bernardino Mountains interacting with ozone released by industries.
   b. nitrogen oxides and hydrocarbons from cars and industries interacting with sunlight.
   c. carbon dioxide and methane that also cause the greenhouse effect.
   d. pollutants from other cities that are blown there and trapped.

2. Ozone, a component of smog, is a particular concern in Los Angeles because
   a. it is also believed to contribute to the greenhouse effect.
   b. the industries that release it have never been regulated.
   c. it causes the most obvious breathing problems and other health-damaging effects.
   d. people fear that loss of ozone in Los Angeles will result in more cancers.
   e. More than one of the above is true.
      The correct answers are _____.

3. Which of the following is *not* an example of output control of pollutants?
   a. three-way catalytic converters on automobiles
   b. sewage treatment plants on tributaries of the Rhine River
   c. the South Coast Air Quality Management District plan
   d. smokestack scrubbers on factories

4. The new concern about airborne toxic chemicals has arisen in Los Angeles because
   a. other, more traditional pollution problems have been largely solved.
   b. their health effects are easy to recognize.
   c. there was a measurable increase of pathogenic occurrences.
   d. the region is one of the few areas in the country where these chemicals are released.
   e. More than one of the above is true. The correct answers are _____.

5. Which of the following is *not* among the pollution control strategies recommended in the proposed South Coast Air Quality Management District plan?
   a. transition to electric-powered cars
   b. incentives for car pooling
   c. expanded reliance on mass transit
   d. a new catalytic converter for automobiles
   e. restrictions on barbecues and other consumer practices

6. The problem of eutrophication in the Rhine River and the North Sea is due to
   a. chemical spills like the Sandoz accident of November 1986.
   b. nutrients from uncontrolled sewage dumped into the river.
   c. nutrients from farmland and other non-point sources throughout the basin.
   d. toxic chemicals including PCBs that have accumulated in Rhine sediments.
   e. the explosion of new growth that has occurred as the river is cleaned up.

7. The catastrophic fire and pesticide spill from a Sandoz chemical warehouse on the Rhine in late 1986
   a. released more chemicals into the river than all industrial sources combined.
   b. had little effect on the river because it had recovered from earlier pollutants.
   c. produced less pollution than the river receives each day from usual sources.
   d. caused an enormous increase in public awareness of pollution of the Rhine.
   e. More than one of the above is true. The correct answers are _____.

8. Water-quality monitoring stations along the Rhine use tests based on living organisms because
   a. fish may reveal the effects of previously unsuspected pollutants.
   b. fish are more sensitive to familiar pollutants than is high-tech laboratory equipment.
   c. fish are cheaper to use than laboratory equipment.
   d. diseases that show up in fish are likely to show up in people.
   e. More than one of the above is true. The correct answers are _____.

9. Bentazone and other micropollutants in Rhine water
   a. are dumped in the river because the companies that produce them consider them safe.
   b. are secondary pollutants formed by unpredictable reactions between more familiar pollutants such as nitrates and phosphates.
   c. are increasing faster than they can be monitored by scientists.
   d. can be clearly linked to human health problems.
   e. More than one of the above is true. The correct answers are _____.

True or False

1. Acid deposition is not a problem in Los Angeles, because there are few industrial sources of sulfur oxides. _____

2. Although the health-damaging effect of air pollutants is well known, damage to trees and vegetation in the Los Angeles basin has been observed only recently. _____

3. People in Los Angeles become physically more tolerant to pollution exposure. _____

4. The major source of smog-causing pollutants in Los Angeles is automobiles. _____

5. The South Coast Air Quality Management District plan is based on approaches that have proved successful in other U.S. cities. _____

Sample Essay Questions

1. Explain the distinction between *primary* and *secondary* pollutants by describing how photochemical smog is formed. Is the ozone in Los Angeles' air a primary or a secondary pollutant?

2. Use an example from "Do We Really Want to Live This Way?" to explain how *output* pollution control methods can be undermined by continued population and economic growth.

3. Define "bioaccumulation" of toxic materials, and explain why this process poses a risk to human health.

4. Describe some of the techniques a risk assessment specialist might use to evaluate the hazard of pollution by the pesticide bentazone in Amsterdam's drinking water supply.

5. Summarize the South Coast Air Quality Management District's plan to reduce air pollution in the Los Angeles basin. Explain whether the plan relies on input controls or output controls. Do you consider the plan realistic? Why or why not?

GET INVOLVED

References

Covelo, V. T., et al. *Risk Evaluation and Management.* New York: Plenum, 1986.

Graedel, Thomas E., and Paul J. Crutzen. "The Changing Atmosphere." *Scientific American*, Sept. 1989.

Krimsky, Sheldon, and Alonzo Plough. *Environmental Hazards: Communicating Risks as a Social Process.* Dover, Mass.: Auburn House, 1988.

MacKenzie, James J., and Mohamed T. El-Ashry, eds. *Air Pollution's Toll on Forests and Crops.* New Haven, Conn.: Yale University Press, 1990.

Maurits la Riviere, J. W. "Threats to the World's Water." *Scientific American*, Sept. 1989.

National Academy of Sciences. *Air Pollution, the Automobile, and Human Health.* Washington, D.C.: National Academy Press, 1988.

National Academy of Sciences. *Drinking Water and Health.* Washington, D.C.: National Academy Press, 1986.

Organization for Economic Cooperation and Development (OECD). *Water Pollution by Fertilizers and Pesticides.* Washington, D.C.: OECD, 1986.

Rodericks, Joseph V. *Calculated Risks: Understanding the Toxicity and Human Health Risks of Chemicals in the Environment.* New York: Cambridge University Press, 1992.

Wagner, Travis. *In Our Backyard: Understanding Pollution and Its Effects.* New York: Van Nostrand Reinhold, 1993.

Several chapters in *Taking Sides: Clashing Views on Controversial Environmental Issues* deal with topics covered in this unit:

- Issue 4. Does Risk-Benefit Analysis Provide an Objective Method for Making Environmental Decisions? (pp. 52–69)

- Issue 7. Should Regulation of Airborne Toxins Under the Clean Air Act Be Strengthened? (pp. 112–125)

- Issue 10. Is There a Cancer Epidemic Due to Industrial Chemicals in the Environment? (pp. 164–183)

Also see chapters dealing with air and water pollution in various editions of Worldwatch Institute's annual *State of the World* report: *State of the World 1990.*

- Anne Platt McGinn, "Phasing Out Persistent Organic Pollutants," *State of the World 2000.*

- Molly O'Meara, "Exploring a New Vision for Cities," *State of the World 1999.*

- Anne E. Platt, "Confronting Infectious Diseases," *State of the World 1996.*

- Marcia D. Lowe, "Rethinking Urban Transport," *State of the World 1991.*

- Hilary French, "Clearing the Air," *State of the World 1990.*

- Sandra Postel, "Protecting Forests from Air Pollution and Acid Rain," *State of the World 1985.*

Organizations

Most environmental groups in the United States and Europe are involved with air and water pollution issues; in many cases these issues first attracted widespread public attention to environmental causes. The groups listed below are involved both with public education and with advocacy of new policies to combat pollution problems such as those discussed in "Do We Really Want to Live This Way?"

Greenpeace USA
1436 U Street NW
Washington, D.C. 20009
Tel: (202) 462–4507; web: www.greenpeaceusa.org

National Toxics Campaign
37 Temple Place
4th Floor
Boston, MA 02111
Tel: (617) 482–1477

Natural Resources Defense Council
1350 New York Avenue NW
Suite 300
Washington, D.C. 20005
Tel: (202) 783–7800; web: www.nrdc.org

UNIT 6

In the Name
of Progress

In the Brazilian rain forest, a *Race to Save the Planet* camera crew documents the Greater Carajas project. This government-sponsored program involves timbering the rain forest and baking logs in primitive ovens to produce some 80,000 tons of charcoal each week, used to smelt iron.

BEFORE YOU VIEW THE TELEVISION PROGRAM

Learning Objectives

After you complete the assigned readings and view "In the Name of Progress," you should be able to

- explain the process of industrialization and describe some of its environmental and social consequences.

- define *sustainable development*, contrast it with conventional views of development, and cite examples of sustainable economic activities.

- explain the relationship between population growth and economic development.

- describe the United States' stake in the development paths chosen by other countries and explain how the United States is involved in their development.

- support your own judgment of whether or not sustainable development is feasible.

Reading Assignment

Choose the material from either textbook as your reading assignment. Your instructor might assign additional readings as well. You might also find the references listed at the end of this Study Guide unit useful.

Living in the Environment

Chapter 11, "Population Dynamics: Influencing Population Size," Sections 11-3 and 11-4, and Guest Essay by Garrett Hardin

Chapter 24, "Sustaining Ecosystems: Deforestation, Biodiversity, and Forest Management"

Chapter 27, "Economics and Environment," Section 27-4

Environmental Science

Chapter 2, "Economics, Politics, Ethics, and Sustainability," Section 2-4

Chapter 9, "The Human Population," Opening, p. 202, Sections 9-3 and 9-4, and Guest Essay by Garrett Hardin

Chapter 17, "Sustaining Terrestrial Ecosystems:

Forests, Rangelands, and Parks," Sections 17-1 through 17-4

Unit Overview

This unit explores the environmental and social dimensions of poverty, the condition in which two-thirds of the world's people live, and explores the various processes of "development" by which nations attempt to achieve material well-being and economic success. Both poverty and development entail profound environmental consequences. In this unit we consider the idea of "sustainable development" in which people's legitimate aspirations are met without damaging the environment or undermining the resource base needed to sustain future generations.

The assigned readings take a close look at population growth and its consequences. The text explains how and why populations tend to increase and why growth rates in developing countries tend to exceed those in industrial countries. Methods of family planning, and the politics surrounding birth control, are also discussed.

The accompanying television program, "In the Name of Progress," compares conventional development and its alternatives in two very different societies, Brazil and India. After reviewing some of the unintended consequences of industrialization in the two societies, the program profiles recent grassroots (local community) efforts to meet the needs of rural communities without environmental destruction. India's Chipko movement and the rubber tappers of Brazil's Amazon region show what sustainable development, within the limits set by the local environment, can look like. The program explores the complex connections among poverty, population growth, and development and looks at the changing role of multi-national institutions such as the World Bank that shape development policies around the world.

Poverty and overdevelopment, two poles of human experience, are both responsible for the global changes that have put the stability of the planet at risk. Throughout later units of the course, we will consider problems and solutions from the standpoint of both affluent and poor societies and appreciate the fact that, despite the deep social and economic gaps dividing its members, the human species shares one biosphere in which the challenges of equity and environmental sustainability must inevitably be met.

Glossary of Key Terms and Concepts

The following terms and concepts will be useful as background for viewing "In the Name of Progress."

The **Chipko movement** is a grassroots social movement begun in the forested foothills of the Himalaya Mountains in Uttar Pradesh, India, in 1973. Chipko, which means "hug" or "cling to" in the Garhwali language of the region, was begun by women who literally clung to trees that commercial loggers planned to cut on their hillsides. The movement has evolved from forest protection to what its members call "ecodevelopment" (see below): small-scale tree-planting and soil conservation activities that restore the environment and benefit local communities.

Contraceptives are devices and drugs that prevent conception. Widely used in industrial countries to allow couples to decide when and how often to have children, modern contraceptives are still far from common in most developing countries where birth rates are highest. The availability of contraceptives is one factor influencing rates of population growth.

Development is a process of economic and social transformation that defies simple definition. Though often viewed as a strictly economic process involving growth and diversification of a country's economy, development entails complex social, cultural, and environmental changes. There are many models of what "development" should look like and many different standards of what constitutes "success."

The **"ecodevelopment camps"** run by leaders of India's Chipko movement (see above) help rural communities in the Himalayan foothills to undertake village-based activities to sustain and improve the local environment by tree planting, soil and water conservation, and small-scale harvesting of natural products. Women are encouraged to take leadership roles in the camps.

A country's **external debt** is the amount of money it has borrowed from foreign creditors, generally from banks or governments in Europe, North America, and Japan, to pay for its development. The interest on such debt must be paid in the currency in which the loan was made, which means that a country must export goods to the country it has borrowed from in order to earn the foreign exchange it needs to pay interest and principal. The "debt crisis" arose when many countries could no longer meet interest payments or could only do so at an unacceptably high social cost within their societies.

"Extractive reserves" are areas of forest designated by the government of Brazil as protected reserves where rubber tappers (see below) can practice their livelihood based on gathering natural forest products without the threat of deforestation by farmers or ranchers. The first such reserves were designated in the western Amazon in the mid-1980s after years of organizing by rubber tappers in the region and support from international environmental groups.

Most governments and world leaders now endorse their people's right to **family planning**—access to the information, contraceptives, and medical care that allow couples to choose the timing and number of their children. Many governments, however, have yet to devote sufficient resources to the task. Access to means of family planning is one of the key determinants of population growth rates.

Mohandas K. **Gandhi** (1869–1948) was a Hindu lawyer who sparked and led the nationwide movement of nonviolent resistance to British colonial administration that gained India's independence from the British empire in 1947. Known as Mahatma ("great soul") to his followers, Gandhi advocated a development path for India based on village-scale cottage industries and programs aimed at meeting basic human needs to alleviate India's pervasive poverty. He was assassinated in 1948.

The **gross national product (GNP)** is a statistical measure of the total economic value of all the goods and services produced within a nation in a given year. The size and rate of growth of GNP are often taken as indicators of the level of development achieved by a society, although GNP contains many items (spending to clean up environmental damage, treat drug addicts, keep criminals in jail, etc.) that reflect social difficulties rather than social well-being.

Industrialization is a development path based on expanding a country's capacity to process raw materials and manufacture products for consumers, businesses, and export. This approach to development, first seen in northern Europe in the Industrial Revolution (see Unit 3), entails heavy financial investments in factories and power plants and a rapidly growing demand for energy, particularly fossil fuels.

Land reform is a means of promoting development in rural areas by correcting inequities in ownership of land and enabling more of the rural population to own the land they farm. Productive land is often owned by a disproportionately small share of a country's population, a situation that is particularly pronounced in Latin America, while large numbers of people lead a marginal existence as tenant farmers. Correcting such inequities through land reform is inevitably a politically volatile and often violent process.

"Modernization" is the imprecise title given to the transformation of a nation or region from an agricultural economy largely aimed at meeting subsistence needs and sustaining low living standards to a more diverse economy in which industry accounts for a considerable share of GNP. Science and technology influence every sphere of economic activity, and living standards (as measured by infant mortality, life expectancy, literacy rates, income levels, etc.) improve. Modernization and industrialization have been viewed as synonymous by many economists and political leaders.

Jawaharlal **Nehru**, India's first prime minister after independence in 1947, believed the country's development could only be accomplished through vigorous nationwide industrialization, essentially a "trickle-down" approach (see below). Large-scale mining, manufacturing, and power-generating facilities lay at the center of Nehru's vision for India.

People whose homes or communities are displaced by large-scale industrial projects such as dams or power plants are called **"oustees"** in India. Though they generally must receive some compensation to move and set up homes in a new area, the disruption of their lives and livelihoods makes them the victims of industrial development.

Per capita income is an economic statistic calculated by dividing a country's gross national product (GNP) by the number of people in its population. Though it bears no direct relation to average levels of income earned by workers within a society, it reflects the amount of wealth available within the society to support each person's needs. This measure is widely used to make international comparisons of the levels of development attained by different societies.

India's oustees, displaced by industrial projects, are often relocated to **resettlement villages** built by the government or by the industry responsible for the project. Though such villages represent a considerable investment beyond the costs of the project itself, the displaced families often find their new lives plagued by insufficient services, inadequate jobs, and unfamiliar social problems.

Rubber tappers live and work in the rain forests of Brazil's Amazon basin, earning their livelihood by tapping latex from natural rubber trees and gathering other forest products, such as Brazil nuts, for sale. Most are descendants of laborers who worked for the "rubber barons" who controlled production in the region's forests in the late nineteenth century. Threatened by settlers seeking to clear the forests for ranches and farms, the rubber tappers have organized a national union that uses nonviolent methods to defend their forests and to prove their livelihoods more sustainable and economically viable than the alternatives involving forest destruction.

A new concept of **sustainable development** has emerged in recent years, based on the premise that development must meet the needs of the present generation without compromising the ability of future generations to meet their needs. Environmental protection and management is central to sustainable development, rather than a luxury that must be deferred until the generation of wealth has made pollution control and other forms of environmental protection "affordable."

Thermal power plants generate electricity by using steam to turn turbines, relying on either fossil fuels or nuclear fission (see Unit 8) to boil the necessary water. Used to provide power to industries and households, they are a keystone of industrialization and among the most costly investments (ranging from hundreds of millions to billions of dollars) made by any society. In developing countries, construction of thermal power plants is often financed with external debt (see above), on the premise that the economic activity made possible by expanded electricity supplies will enable the country to pay off its loans.

According to the **"trickle-down theory"** of development, growth in a country's GNP and per capita income (see above) made possible by large-scale industrialization would inevitably improve the living standards of all of a country's citizens through creation of jobs and expanded economic opportunities. Widely prevalent in the 1950s, this view of development as a purely economic phenomenon influenced government policies in many countries and the lending programs of many creditors such as the World Bank (see below).

The **World Bank**, officially called the International Bank for Reconstruction and Development, is a multinational financial institution created by the United States and its European allies at the end of World War II to finance the reconstruction of Europe. Its mission was soon broadened to make investments in agriculture and industrial development in the poorer nations of Africa, Asia, and Latin America. In recent years, environmentalists have criticized the Bank for supporting projects that destroy irreplaceable natural resources (for example, financing roads that have hastened destruction of Amazon forests), resulting in some changes in its lending policy and the creation of an Environment Department.

AFTER YOU VIEW THE TELEVISION PROGRAM

Consider What You Have Seen

"In the Name of Progress" poses many questions about the goals of development and the means chosen to achieve them. As you think about the material presented in the program, take time to reflect on your own life: What is essential to your sense of well-being? What must be done to supply or sustain the standard of living you enjoy? The following themes may help you relate the program to the reading assignment.

- What is development?
- Population and development
- Energy: A double-edged sword
- Industrialization and its consequences
- Models of sustainable development

What Is Development?

The "dreams" of development: This program opens with a discussion of whether the dreams of development in India are the same as those of the West. This tells us that development is not an abstract process, but rather something guided by people's aspirations and hopes for the future. Think about the various "dreams" shown in the program. How do the dreams of the operator of Hindalco Aluminum, part of the industrial complex at Singrauli, India, differ from those of the people of Kachni village, next in line to be displaced by the complex? How do the aspirations of Chandi Prasad Bhatt, leader of the Chipko movement in the Himalayas, differ from those of the workers who cut trees for the pig-iron smelters at Brazil's Greater Carajas complex? What do these dreams, or visions of the future, have in common? How do they differ?

An economic process: In its most conventional definition, development is a process of economic change that results in modernization, or lifting living standards to a level comparable with those achieved in Europe or North America. That has historically involved developing industries, building power plants and roads, and shifting from a farming-based economy to one that relies on manufacturing and is dominated by cities. All of these changes affect how a majority of people make their living and what they hope will be possible for their children. All of these changes alter the relationship between humans and their environment.

The various examples shown in the television program involve changes in people's livelihoods, but not all fit the profile of modernization summarized above. Think about the Chipko movement in the Himalayas and the rubber tappers of Brazil.

Do you think these popular movements fit the conventional model of development, or do they represent a different sort of process?

The debt crisis: Conventional development involves mobilizing large sums of money to construct the factories, dams, and power plants needed to expand a country's GNP. Almost all economic change involves loans or credit; the earnings from a new enterprise enable a borrower to pay back the loan with interest. To finance their development projects, governments also borrow money both from wealthier governments and from commercial banks. In recent years, poor countries' inability to pay the interest charges on their international borrowings has led to the debt crisis.

Brazil, as shown in the program, owes $120 billion to foreign creditors, more than any other developing country owes. Although the money was borrowed to make the country more productive, the burden of massive debt now undermines social welfare and accelerates environmental damage in Brazil and many other countries. For Brazil, like many developing nations, farm products, fisheries, and timber are the easiest products to export in order to meet interest payments, often at a cost of deforestation, soil erosion, overfishing, and other effects that reduce the capacity of the environment to sustain a harvest in the long run. Thus, Third World debt is not just a financial matter, but a human crisis and an environmental crisis.

While you are enrolled in this course, follow news about the debt crisis in your local newspaper or a national newspaper such as the *New York Times*. Keep a list of the countries shown in the television programs that have problems with foreign debt. What plans are being considered to deal with the debt crisis? What do you think their environmental effects might be?

Who makes development happen?   The program shows a number of institutions involved in setting development priorities. The World Bank, shown in the program (see the glossary), makes loans to developing countries to establish industries, build roads, improve agriculture, and otherwise hasten the process of modernization. The World Bank is influential, as the program suggests. But it is not the *only* institution that funds development, or even the biggest. Governments make loans to one another, foreign aid agencies grant money for development,

and thousands of nongovernmental organizations in both industrial and developing countries support development projects. The largest amount of money is that spent by governments within their own borders.

The World Bank is singled out in the program because it is a public institution, supported by governments and responsive to public pressure—such as the environmental activists shown gathered in Berlin. But at the Singrauli industrial complex in India and the Carajas mining complex in Brazil, the most important decisions are made by national authorities, not international institutions. How much influence do you think North Americans should have over decisions made in other countries? Do you think Americans should be concerned about development choices made in other parts of the world?

Nehru vs. Gandhi: The contrasting visions of India's Prime Minister Nehru and Mahatma Gandhi, presented in the program, express very different ideas of what development means. Which vision of development—Nehru's or Gandhi's—would you associate with each of the examples shown in the film: the Singrauli industrial complex, the Chipko movement, the Carajas mining complex, the rubber tapper movement, and the World Bank? Which vision do you favor? Do you think the two are mutually exclusive?

Population and Development

The logic of rapid population growth: Your reading assignment introduces the dynamics of population growth and examines why developing countries tend to grow more rapidly than industrial countries. "In the Name of Progress" explores the logic of population growth in a village called Masana in rural India. List some of the reasons mentioned as responsible for large families in Masana. As you reflect on your readings, what factors do you think the program fails to mention?

Development: The best contraceptive?   Recall that the Indian researcher interviewed in the television program concludes that "development is the best contraceptive." Paraphrase this statement in your own words. The reading assignment for this unit discusses the concept of the "demographic transition." Review this concept, and decide whether it relates to the case of Masana village shown in the television program.

Both the villages involved in the Chipko movement in the Himalayas and the rubber tapper communities in Brazil's Amazon basin are pursuing a development path quite different from the conventional model of modernization assumed by demographic transition theory. What effect, if any, do you think these movements might have on rates of population growth? What effect do you think the prominent role of women in the Chipko movement might have on desired numbers of children in the community?

Country comparisons: Brazil and India, though both among the five most populous countries on earth, are quite different in size. In 1999, Brazil had 168 million people, while India's population was estimated at 987 million. India's population was expanding at about 1.9% per year and Brazil's at 1.5%. At this rate, calculate by how many millions of people India's *annual* growth exceeds Brazil's.

Review the concepts of exponential growth and population doubling times, introduced in the reading assignment. By what date would you expect populations in the two featured countries to have doubled in size if growth rates remain at present levels?

Personal choices: Whether at the level of a local community, a nation, or the entire planet, the growth of the human population is a result of individual choices about reproduction and family size. List the factors that influence your own feelings about family size. How important are economic factors? Do you want (or have) more or fewer children than your parents had?

Energy: A Double-Edged Sword

Why development requires energy: All economic and subsistence activities require energy. Energy is needed to fabricate products from raw material, to transport goods and people, and to provide amenities such as light and refrigeration that are associated with modernization. As you learned in Unit 3, major economic and social changes (such as the Industrial Revolution) nearly always go hand-in-hand with changes in the energy resources used by society. List the sources of energy used by the Singrauli thermal power complex and by the Carajas mining and smelting operation. What sources of

energy are used by the people in Bemeru village in the Himalayas (Chipko movement) or the community of Xapuri in the Brazilian Amazon (rubber tappers)? How do these energy sources differ from those used by large industrial development projects?

Questions of scale: Large-scale energy use invariably causes large-scale environmental consequences. What are some of the environmental impacts associated with large-scale industrial development in Brazil and India? What steps to address these impacts are mentioned in the television program?

"Free" energy brings environmental destruction: The fuel source for the pig-iron smelters at the Greater Carajas complex in Brazil is charcoal, produced from timber in the surrounding rain forest. Recall why the workers at Carajas cut trees without replanting. Much of the iron ore produced at Carajas is exported to help Brazil pay off its foreign debt. Why do you think Brazilian authorities might be reluctant to replant the forests cut for charcoal or to raise tree plantations for the smelters instead of cutting virgin forest?

Industrialization and Its Consequences

Environmental consequences: Make a list of some of the environmental consequences of the various industrial development projects shown in "In the Name of Progress." Industrial projects have both costs and benefits. Do you think the program provides a balanced view of both?

Social consequences: Industrial development invariably changes relationships among people in society, producing both "winners" and "losers." Who are the winners and losers in the Singrauli and Carajas schemes? Do you think both groups were considered by the planners of each project? Could either project have been developed without its negative or disruptive effects on neighboring communities?

"Overdevelopment": Think back on the material presented in Unit 5, "Do We Really Want to Live This Way?" How do you think the advocates of industrial development in India or Brazil shown in this unit might respond to the environmental problems that have developed in the Los Angeles basin or the Rhine River valley? Do you think the authorities coping with the effects of overdevelopment in those places would encourage planners in developing countries to give more or less weight to environmental concerns at the early stages of industrial development?

Models of Sustainable Development

Questions of scale: At the end of "In the Name of Progress," the narrator says that "development is meaningless without protection of the resources on which that development lies." The program shows two models of sustainable development that meet this criterion: the Chipko movement in the northern Himalayas and the rubber tapper organization in the Amazon basin. Could either model be adopted by an entire society the size of Brazil or India?

Pattern of resource use: What resources are used by Chipko communities and by rubber tappers? How does the pattern of resource use in each case demonstrate the principle mentioned above, "protection of the resources on which development lies"?

The population factor: The communities involved in these two experiments in sustainable devel-opment are remote rural communities, in which one might expect that cultural attitudes and economic imperatives would favor large families and rapid population growth. Do you think that either the Chipko or the rubber tapper approach to development would change the logic favoring large families? How? How might population growth affect the sustainability—and the environmental impact—of these two examples?

What are the limits?   How do you think David Hopper, the vice president of the World Bank shown in "In the Name of Progress," might respond to the Chipko and rubber tapper examples? What do these development alternatives offer that the more conventional view of development through industrialization leaves out? After having viewed the television program, what do you think it would take to make industrial development sustainable?

Take a Closer Look at the Featured Countries

Brazil and India, two giants among developing countries, together are home to nearly one-fifth of the earth's population. Despite their extraordinary natural and cultural diversity, these two countries share common economic and environmental problems that shape their development prospects. Though no developing country can be said to be "typical," Brazil and India both confront problems faced by millions of people in societies all over the world who are struggling to overcome their poverty.

Brazil's population of 168 million (1999) occupies the fifth largest country on earth (Figure 6-1). The population is currently expanding by 1.5%—2.5 million people—each year. As in many other developing countries, a large share of Brazil's population—32%, or 54 million people—is under 15 years of age. The Amazon Basin, which covers about one-third of the country's territory, remains sparsely settled compared to eastern and southern Brazil, and the roads and rivers penetrating Amazon forests have long been treated as a demographic "safety valve" for the more-crowded coastal areas. While São Paulo, Rio de Janeiro, and other large cities are as cosmopolitan as any on earth, Brazil's sprawling urban slums and unrelieved rural poverty have led some to describe each of these communities as "a Belgium within an India"—an island of modern industrial society in a sea of entrenched rural poverty. The enormous gulf that separates rich and poor in Brazil lies at the root of many of the country's most serious environmental problems.

In 1997, Brazil's per capita gross national product (GNP) totaled $4,790 per person, compared to the United States' per capita GNP of about $29,080 per person. According to the World Bank, this level of income classifies Brazil as a "middle-income" country. With a diverse economy, Brazil is not a major recipient of foreign aid; official development assistance in 1991 accounted for only 0.1% of the country's GNP. However, Brazil has fueled much of its recent growth by borrowing abroad, and the country's total external debt, estimated   at $111   billion   in   1986,

Figure 6-1   Brazil

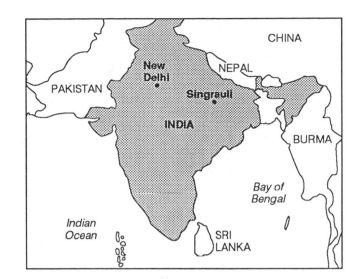

Figure 6-2   India

was the largest of any developing country. Of every $100 Brazil earned from its exports of soybeans, beef, minerals, and manufactured products, $42 was paid out in interest to the country's creditors. This heavy debt burden increases pressures on Brazil's resource base as the country struggles to expand exports at all costs.

The cutting and burning of the once vast forests of the Amazon basin, perhaps Brazil's most notorious environmental problem, is only partly driven by these economic pressures. Brazil's leaders and many of its citizens think of the Amazon much as Americans regarded the Great Plains a century ago: a frontier to be conquered and tamed. Cattle ranching has been particularly important in the "conquest" of this frontier, aided by generous government incentives, and the clearing of pastures is the single greatest cause of deforestation. At recent count (1984–86), Brazil had 127.6 million head of cattle, nearly one cow per person. By contrast, the country had about 10 million cars, or one for each 15 Brazilians.

Brazil's leaders have staked the country's future on an industrial model, typified by large-scale projects like the Greater Carajas project. It is this path that accounts for much of the debt that now weighs so heavily on Brazil's future. But recent years have seen the emergence of different visions for the future of the country, as groups like the rubber tappers and many tribal peoples native to Amazon forests have organized to protect their livelihoods and the environment that supports them. Whether development alternatives like these will remain isolated cases or shape a new philosophy of development equal to the challenge

of the country's perilous financial status is uncertain, but the debate over what constitutes sustainable development for Brazil is now in full swing.

With 987 million people, **India** ranks after China as the world's second most populous nation (Figure 6-2). India's population is growing by 1.9% annually; in a country of India's size, this rate of expansion means 19 million more Indians each year. India is expected to overtake China as the world's population giant by early in the twenty-first century. Sheer size masks the extraordinary diversity of India's people; India has four major religions, and in addition to the official languages Hindi and English, a dozen other regional languages are recognized by the country's constitution. Three-quarters of India's people live in rural villages, and 36%, 355 million people, are children younger than 15 years. The United Nations estimated that only 57% of Indians had access to clean drinking water in 1985 and less than 10% had access to sanitation. Not surprisingly, poverty is endemic and entrenched in India, in both the cities and the rural areas. More than one child dies in infancy for every ten born, perhaps the most tragic indicator of deep poverty.

In 1997 India's average GNP per person was only $370, a statistical confirmation of the country's poverty. The World Bank classifies India as a "low-income" country. Foreign aid totaled just over $2 billion in 1991, about 0.7% of India's GNP—a small part of the country's total economy. Like other nations, India has borrowed heavily to pay for its modernization. In 1986, India owed $41 billion to creditors, and 27% of the country's export earnings went to pay interest on this debt. Relatively few Indians have access to the conveniences of the modern world. For example, in 1986 there were just 1.5 million cars in the entire country, or one for every 544 people. The living standards taken for granted in industrial societies are almost certain to remain out of reach for the vast bulk of the Indian population.

With three-quarters of the country's people classified as "rural," a central challenge of development in India is how to manage farmlands, forests, and other living resources on a sustainable basis. Indian environmentalist Anil Agarwal has called for attention to India's "gross nature product," the capacity of the natural environment to supply the food, firewood, and fodder needed for decent livelihoods. Environmental degradation in many forms—deforestation, spreading deserts, the abandonment of once-productive farmland—is chronic in many parts of the country, but some communities (such as those in the Garhwal Himalayas shown in the television program) are struggling to create a future based on careful stewardship of their resources. Without attention to the forces driving population growth throughout the country, the few shining examples of ecodevelopment may have come too late to make a difference to India's future.

Examine Your Views and Values

1. Make a list of your impressions of developing countries. Which of the following sources of information have most influenced your impressions: personal experience, family members, friends or acquaintances, television, newspapers, classes? Do you think that the dreams of the developing world are different than those of the West?

2. The U.S. population of 263 million people is growing by about 0.7%, or 1.8 million people, each year. Do you think the United States should have an official policy on population growth? Why or why not? What should that policy be? Should the United States support

efforts to restrain population growth in other nations? Why or why not?

3. The television program draws a distinction between "developing" countries such as Brazil and India and "industrialized" countries such as the United States. Can you think of ways in which the United States is still "developing"? Is any society ever completely "developed"?

TEST YOUR COMPREHENSION

Self-Test Questions
(*Answers at end of Study Guide*)

Multiple Choice

1. With projects like the Singrauli thermal power complex, India has chosen a development path
   a. completely different from that undertaken by countries in Europe and North America.
   b. very similar to the model of development followed by countries in Europe and North America.
   c. in which industrialization plays a minor role.
   d. in which industrialization plays a major role.
   e. More than one of the above is true.
      The correct answers are _____.

2. Gandhi felt that India's development
   a. must begin with the impoverished families in India's 700,000 villages.
   b. should be based on both large-scale industrialization and meeting human needs.
   c. could be based on small-scale industries that supplied local people with essentials.
   d. would require assistance from industrial countries such as Great Britain and the United States.
   e. More than one of the above is true.
      The correct answers are _____.

3. The families in communities adjacent to Singrauli
   a. were among the first to benefit because of their proximity to power supplies.
   b. long ago stopped using wood for fuel because electricity is a much more efficient fuel source.
   c. have been moved to resettlement villages because their land was needed for the complex.
   d. have been moved to resettlement villages because they chose a new location for their homes.
   e. More than one of the above is true.
      The correct answers are _____.

4. The Indian environmentalist Smitu Kothari, who visits Singrauli in the television program, believes that
   a. the Singrauli complex proves that the trickle-down theory of development doesn't work in India.
   b. the Singrauli complex, despite its unfortunate effects, still shows the benefits of industrialization.
   c. the Singrauli complex could have been more beneficial if it had been managed by environmentalists.
   d. the costs and benefits of Singrauli just about cancel each other out.
   e. More than one of the above is true.
      The correct answers are _____.

5. The region in which the Chipko movement began
   a. is so remote that it has few environmental problems.
   b. is free of the problem of population growth that plagues other parts of India.
   c. is steep and prone to soil erosion.
   d. is home to tribal people who find it difficult to work together.
   e. More than one of the above is true.
      The correct answers are _____.

6. Women play a unique role in Chipko because
   a. men don't have time to work on environmental projects.
   b. women's activities are directly dependent on the health of forests and the environment.
   c. women are leaders of Chipko as well as followers.
   d. men are not allowed to participate in the movement.
   e. More than one of the above is true.
      The correct answers are _____.

7. Because of its population size and rate of growth, India
   a. is already the most populous nation on earth.
   b. may be the most populous nation on earth in a century.
   c. may be the most populous nation on earth by the turn of the century.
   d. will soon stabilize and begin to shrink in size.
   e. More than one of the above is true.
      The correct answers are _____.

8. Rural families in villages like Masana choose to have more children because
   a. children bring income and security to their parents.
   b. children can help with farm work.
   c. population growth is not a problem in India's rural areas, unlike the cities.
   d. new farming techniques have made most families richer and thus able to support more children.
   e. More than one of the above is true.
      The correct answers are _____.

9. The television program suggests that Brazil's development has not yet been successful by showing that
   a. the largest slum in Latin America is in Rio de Janeiro.
   b. the rain forests of the Amazon basin are being destroyed.
   c. children in Brazil still die of preventable diseases such as measles.
   d. Brazil has the largest external debt in Latin America.
   e. More than one of the above is true.
      The correct answers are _____.

10. The Greater Carajas project in the eastern Amazon
    a. is a huge power complex designed to supply Brazil with electricity.
    b. is an industrial development scheme that includes iron mines, sawmills, ranches, and plantations.
    c. is a project the World Bank considers a showpiece of sustainable development.
    d. is a model of ecodevelopment inspired by local communities.
    e. More than one of the above is true.
       The correct answers are _____.

11. Satellite pictures show that deforestation in the Carajas region
    a. has been exaggerated by environmentalists.
    b. could lead to complete deforestation by 2010.
    c. was rapid at the start of the project but has since tapered off.
    d. looks serious on the ground but cannot be detected from space.
    e. More than one of the above is true.
       The correct answers are _____.

12. Chico Mendes founded the National Union of Rubber Tappers because
    a. the rubber tappers' livelihood was threatened by ranchers, farmers, and miners invading their forest areas.
    b. powerful landowners used violence to intimidate the rubber tappers.
    c. environmentalists convinced him that the forests needed protection.
    d. the rubber tappers' livelihood could not be justified on economic grounds alone.
    e. More than one of the above is true.
       The correct answers are _____.

Fill In

1. The thermal power complex at Singrauli poses a number of environmental problems. Although the coal-fired power plants are equipped with _____ , these control only _____ , not _____ or _____ . Local water supplies and adjacent farmland have also been contaminated by _____ from the plant.

2. The grassroots movement based on protecting forests that arose in the Himalayas is called _____ . Its leaders were inspired by Gandhi's example of _____ . The forests that the movement's followers sought to protect supplied them with _____ , _____ , and _____ .

3. The people who gather latex in the western Amazon region are known as _____ . Unlike other forms of development in the Amazon, their livelihood does not cause _____ and is considered by many to be a model of _____ . Their work is threatened by _____ , _____ , and _____ who come to the Amazon in search of land.

Matching

Select the letter or letters in Column B most closely associated with each name or place listed in Column A.

Column A	Column B
1. Prime Minister Nehru _____	a. ecodevelopment camps
2. Mahatma Gandhi _____	b. industrialization
3. Singrauli, India _____	c. sustainable development
4. Chandi Prasad Bhatt _____	d. trickle-down development
5. Chico Mendes _____	e. oustees
6. World Bank _____	f. resettlement villages
7. Amazon basin _____	g. land conflicts
8. Greater Carajas project _____	h. pollution and toxic ash
9. Chipko movement _____	i. nonviolent protest
	j. women's participation
	k. forest protection
	l. extractive reserves
	m. forest destruction
	n. short-term perspective
	o. rubber tappers' union

Sample Essay Questions

1. Compare and contrast the environmental problems associated with *industrial* development in Brazil and India, using examples shown in "In the Name of Progress."

2. Describe the Chipko movement, and list the advantages and disadvantages of the development model its followers advocate.

3. Describe the development model advocated by Brazil's rubber tappers, and list its strengths and limitations.

4. Explain how development has encouraged deforestation in Brazil, and tell what might be done to address the problem.

5. Discuss the statement "development is the best contraceptive" with reference to the Singrauli industrial complex in India and the Greater Carajas mining project in Brazil, as shown in "In the Name of Progress."

GET INVOLVED

References

Daly, Herman E. *Beyond Growth: The Economics of Sustainable Development*. Boston: Beacon, 1996.

Hawken, Paul. *The Ecology of Commerce*. New York: Harper-Collins, 1994.

Hawken, Paul, Amory Levine, and Hunter Levine. *Natural Capitalism*. Boston: Little, Brown, 1999.

MacNeill, Jim. "Strategies for Sustainable Economic Development." *Scientific American*, Sept. 1989.

Roodman, Alan M. *The Natural Wealth of Nations: Harnessing the Market for the Environment*. New York: Norton, 1998.

World Commission on Environment and Development. "Towards Sustainable Development." Chapter 2 in *Our Common Future*. (New York: Oxford University Press, 1987).

Also see chapters on development and population topics in various editions of Worldwatch Institute's annual *State of the World* report:

- Lester R. Brown and Christopher Flavin, "A New Economy for a New Century," *State of the World 1999*.

- David Malin Roodman, "Building a Sustainable Society," *State of the World 1999*.

- Lester R. Brown, "The Future of Growth," *State of the World 1998*.

- Lester R. Brown and Jennifer Mitchell, "Building a New Economy," *State of the World 1998*.

- Hilary F. French, "Assessing Private Capital Flows to Developing Countries," *State of the World 1998*.

- Lester R. Brown, "The Acceleration of History," *State of the World 1996*.

- David Malin Roodman, "Harnessing the Market for the Environment," *State of the World 1996*.

- Aaron Sachs, "Upholding Human Rights and Environmental Justice," *State of the World 1996*.

- Lester R. Brown, "Nature's Limits," *State of the World 1995*.

- Hilary F. French, "Forging a New Global Partnership," *State of the World 1995*.

- Sandra Postel and Christopher Flavin, "Reshaping the Global Economy," *State of the World 1991*.

- Alan B. Durning, "Ending Poverty," *State of the World 1990*.

- Alan B. Durning, "Mobilizing at the Grassroots," *State of the World 1989*.

- Jodi L. Jacobson, "Planning the Global Family," *State of the World 1988*.

A number of periodicals are good sources of information and current perspectives on poverty and development issues, particularly the environmental dimension of development. You may find the following in your local library.

Development Forum, a bimonthly newspaper published by the United Nations Department of Public Information, P.O. Box 5850, G.C. P.O., New York, NY 10163-5850

New Internationalist, published monthly in England and Canada and available from P.O. Box 1143, Lewiston, NY 14092

Panoscope, published six times a year by the Panos Institute, 1409 King Street, Alexandria, VA 22314

World Watch, published bimonthly by Worldwatch Institute, 1776 Massachusetts Avenue NW, Washington, DC 20036

Organizations

Hundreds of organizations worldwide are working to correct the inequities that tend to keep poor people poor and to reverse environmental degradation caused by the desperation of poverty. Many groups are pioneering development activities at the local level that resemble the grassroots efforts shown in "In the Name of Progress." The following, chosen to display a range of activities from public education and policy advocacy on development issues to volunteer work in developing countries, are only a tiny sample of the many groups now dedicated to sustainable development.

Environmental Defense Fund
International
257 Park Avenue South
New York, NY 10010
Tel: (212) 505-2100; web: www.edf.org

Global Tomorrow Coalition
1325 G Street NW
Suite 920
Washington, DC 20005
(202) 628-4016

Institute for Food and Development Policy
145 Ninth Street
San Francisco, CA 94103
(415) 864-8555

Oxfam-America
115 Broadway
Boston, MA 02116
(617) 482-1211

World Neighbors
5116 North Portland Avenue
Oklahoma City, OK 73112
(800) 242-6387

PART II

MAKING CHOICES
FOR THE FUTURE

UNIT 7

Remnants of Eden

CHRIS MCCULLOUGH

A member of the biological research team captures a great hornbill in Khao Yai National Park, Thailand. As Thailand's first wildlife park, Khao Yai was established to conserve the country's shrinking forests and to preserve habitat for birds, and animals. However, many villagers on the park's boundaries lost access to resources they depended on when the park was established.

BEFORE YOU VIEW THE TELEVISION PROGRAM

Learning Objectives

After completing the assigned readings and viewing "Remnants of Eden," you should be able to

- define biological diversity and summarize the present concern over the extinction of plant and animal species.

- list the major causes of species extinction due to human activities.

- explain why tropical forests are so important to efforts to preserve the earth's biological diversity.

- cite examples that illustrate why creation of a national park may not be enough to ensure the survival of its characteristic species.

- discuss how the economic needs of local people determine their impact on ecosystems and biological diversity.

- discuss whether you think the area where you live supports its original diversity of living species and describe some of the human impacts that have affected that diversity.

- list the reasons that temperate ecosystems support less biodiversity than tropical ecosystems.

Reading Assignment

Choose the material from either textbook as your reading assignment. Your instructor might assign additional readings as well.

Living in the Environment

Chapter 10, "Population Dynamics," Sections 10-5 and 10-6

Chapter 23, "Sustaining Ecosystems: Land Use, Conservation, and Management," Sections 23-4, 23-6, and 23-7

Chapter 24, "Sustaining Ecosystems: Deforestation, Biodiversity, and Forest Management"

Chapter 25, "Sustaining Wild Species"

Environmental Science

Chapter 5, "Evolution, Biodiversity, and Community Processes," Sections 5-5 through 5-7

Chapter 7, "Population Dynamics, Carrying Capacity, and Conservation Biology," Sections 7-1 and 7-3 through 7-5

Chapter 17, "Sustaining Terrestrial Ecosystems: Forests, Rangelands, Parks, and Wilderness," Sections 17-5 through 17-7

Chapter 18, "Sustaining Wild Species"

Unit Overview

This unit of *Race to Save the Planet* concerns the ultimate foundation of life on earth: the diversity of plant and animal species and the ecosystems those species make up. We will consider how human activities change ecosystems and affect individual species and examine the evidence that the earth is undergoing an extinction crisis unprecedented in life's 3.5-billion-year history. Scientists now recognize that the earth's biological diversity, or **biodiversity**, helps maintain planetary stability and supplies irreplaceable resources. We will look at ways to reconcile human needs with the maintenance of the species and ecosystems on which the planet's future depends.

The assigned readings discuss how populations, communities, and ecosystems respond to stresses such as air pollution and climate change and explain the role of biological diversity in maintaining the resilience of ecosystems. Other assigned chapters discuss land and wild plant and animal species as *resources*, emphasizing human management of forests, wildlife, and fisheries, and introduce the national and international policies enacted to protect wilderness and wild species.

The accompanying television program, "Remnants of Eden," reviews the impacts of human activities on the earth's biological diversity and evaluates different strategies to protect species and wild lands. The program emphasizes the dependence of species on the ecological habitat that supports them and investigates the diversity of species in the earth's richest terrestrial ecosystems, the tropical rain forests. Much of the program explores the uneasy balance between human needs and conservation in national parks in Thailand, Kenya, the United States, and Costa Rica.

Must a human population growing past 6 billion inevitably extinguish species, diminish biodiversity, and disrupt the course of evolution? The question is a crucial one, for both moral and practical reasons. The human species is only one of the 30 million or more that share the planet. As the foundation of the planet's ecological resilience, biodiversity provides an irreplaceable resource on which humans will depend as global changes accelerate. Thus the fate of biodiversity has become a central concern in the race to save the planet.

Glossary of Key Terms and Concepts

The following terms and concepts will be useful as background for viewing "Remnants of Eden." You will recognize some from the material presented in Unit 2.

Biologist Daniel Janzen coined the term **"bio-cultural reserves"** to describe national parks and protected areas that fully involve local people in the management and education activities conducted within them. Janzen believes that the biodiversity protected in parks is, like the artworks and antiquities housed in museums, an invaluable part of national heritage.

Biological diversity (or **biodiversity**) refers to the full array of living species of plants, animals, and microorganisms on earth, as well as the range of genetic variety within each species.

Scientists use **captive breeding** to propagate in captivity some wild animal species that are dwindling toward extinction in their natural habitats. Despite occasional successes like the reintroduction of captive-bred Arabian white oryx shown in "Remnants of Eden," only a small number of wild species can be successfully bred in zoos. It is very difficult to reintroduce a top predator into the wild; most successful reintroductions have involved grazing animals.

An **ecosystem** encompasses the full community of plant and animal species that occur together in a particular place along with their physical and chemical environment.

Species originate and change through a process of **evolution** by which small, random genetic changes allow some individuals to survive better and produce more offspring than others do. The differential survival and reproduction, or natural selection, of individuals best adapted to their habitat at any given time allows species to change as conditions change and accounts for the extraordinary biological diversity of the earth.

A species that cannot reproduce effectively because its habitat has been destroyed or its numbers reduced is destined for **extinction.** The complete disappearance of a species is final and irreversible; species originate only once in the history of life.

Organisms within an ecosystem are linked in **food chains** that describe the sequence of feeding relationships in the system, or "who eats whom." Food chains are rarely separate from one another; chains are linked in intricate networks known as **food webs.**

The **habitat** of a plant or animal species is the place or type of ecosystem in which it is commonly found.

The **Harasis bedouin** are nomadic people native to the arid lands of the Arabian Peninsula whose livelihoods depend on livestock herding. They have assisted the reintroduction of the white oryx to its desert habitat in Oman.

An **indicator species** is a plant or animal species whose presence and abundance reveal the general condition of its habitat. The great hornbill studied in Khao Yai National Park, as shown in "Remnants of Eden," is an indicator species for virgin tropical rain forest in Thailand.

The **Maasai** people are a nomadic cattle-herding tribe native to the savannas (see below) of Kenya and Tanzania in East Africa. Some Maasai have in recent years begun to settle down and practice agriculture.

Minimum critical area refers to the smallest amount of habitat needed to sustain a viable population of some plant or animal species. The concept is used to design national parks large enough to protect rare or unique species.

Poaching means the illegal hunting or gathering of wild species, often for sale on the black market. A number of large mammals, including the African elephant and black rhinoceros, are endangered because of excessive poaching.

The **Population and Community Development Association (PDA)** is a nongovernmental organization in Thailand created to assist local communities with family planning education and small-scale development projects. In "Remnants of Eden," PDA is shown assisting rural communities on the boundary of Khao Yai National Park.

Restoration refers to efforts to reestablish the full community of plants and animals characteristic of a particular ecosystem on land that has been degraded or transformed by human activities.

Savanna ecosystems are grasslands with scattered deciduous trees, found in tropical regions with lengthy dry seasons. Savannas typically support spectacular herds of large hoofed mammals that graze the new growth after seasonal rains.

The **tropical dry forest**, one of the most threatened ecosystem types on earth, covers a tiny portion of its original range in southern Mexico and Central America. Deciduous trees, which dominate this diverse ecosystem, lose their foliage during a dry season.

Tropical rain forests are found near the equator in places where rain falls year-round. The evergreen vegetation and abundant moisture in rain forests sustain the greatest concentration of biodiversity found in any ecosystem type, at least 50% of all species of plants and animals on earth. The rapid destruction of rain forests to make way for human settlement and farming drives thousands of species to extinction.

Wetlands, low-lying areas flooded with shallow fresh or salt water, play irreplaceable ecological roles by purifying water and providing spawning grounds for fish and critical habitat for shellfish, shorebirds, and other species.

AFTER YOU VIEW THE TELEVISION PROGRAM

Consider What You Have Seen

"Remnants of Eden" explores a number of dimensions of human coexistence with biodiversity. The following themes may help you relate what you have seen to the ecosystem concepts introduced in the reading assignment.

- Basic ecosystem principles
- Studying biodiversity
- Beyond park boundaries
- Balancing conservation with local needs
- Undoing the damage

Basic Ecosystem Principles

Five ecosystem types, or **biomes**, are shown in the program: tropical rain forest, tropical dry forest, desert, freshwater lake, and freshwater wetlands. List some of the differences that distinguish these ecosystems. Review the reading assignment for Unit 2, "Understanding Ecosystems," for a more complete list of biomes and their characteristic appearance and species. What type of biome predominates where you live?

All the case studies shown in the program are located in tropical or subtropical regions. Review your text to understand some of the ways that temperate-zone ecosystems differ from those in the tropics. You should be able to list some of the reasons that temperate ecosystems support less biodiversity than tropical ecosystems do.

Many examples in the program illustrate the trophic structure of ecosystems: the pattern of food chains and levels of productivity. The scientists shown studying the fisheries of Lake Naivasha in Kenya are concerned with the lake's food chain and its ability to support a harvest by local fishermen. The Everglades research shows how the character of the entire ecosystem is influenced by which type of tree predominates on the banks of the Kissimmee River's channels. Using what you learned about food chains in Unit 2, why might you expect the replacement of willows by wax myrtles to affect the types and abundance of wading birds found in the channels?

The physical factors that shape ecosystems, though not discussed in detail, are vividly demonstrated in the program. Compare the role of water in the Everglades and in Amboseli National Park. Recall what Daniel Janzen says about the role of fire in tropical dry forests. Recall, too, how the structural complexity of tropical forests is related to the diversity of species those forests contain. Why would you expect species to be lost when such forests are cleared for cropland or pasture?

Studying Biodiversity

Although our lives are surrounded by—embedded in—ecosystems, biodiversity is one of the least known resources on earth. "Remnants of Eden" shows examples of scientists investigating biodiversity in a number of different ways. Think about the techniques used and the kind of knowledge they yield:

- Entomologist Terry Erwin sprays a biodegradable insecticide into the canopy of a tropical rain forest in Peru—a part of the ecosystem that cannot be easily reached by scientists. After collecting the insects that fall to the ground, Erwin painstakingly identifies and catalogs them, searching for new species and unsuspected patterns.

- In Khao Yai National Park, Thailand, Dr. Pilai Poonswad studies great hornbills—a large, easily identifiable species—by tracking them using radio collars. By knowing a great deal about the movements of this one species, Dr. Poonswad and her colleagues gauge the health of the entire forest ecosystem.

- In Lake Naivasha, Kenya, David Harper of Leicester University captures and measures different species of fish to develop a picture of the food chains that provide the structure of this aquatic ecosystem and to see how fishermen's catches might be increased without depleting the lake's productivity.

- In Guanacaste National Park, Costa Rica, biologist Daniel Janzen studies the complex tropical dry forest ecosystem by trying to reassemble it. His chief tool is one of the physical factors that shapes ecosystems: fire.

These examples show just a few of the many ways scientists investigate ecosystems. All these approaches yield crucial knowledge of biodiversity, because no ecosystem on earth is yet thoroughly understood. Much of the work on biodiversity is painstaking, time-consuming, and labor-intensive. And the work has just begun: Of the estimated 30 million or more plant and animal species on earth, scientists have named and described just 1.4 million.

Beyond Park Boundaries

The program shows that parks created to protect biodiversity are invariably affected by factors beyond park boundaries. Unless those forces—economic and political as well as physical—are recognized and managed, biodiversity is likely to be steadily lost.

- The villagers of Ban Sap Tai have traditionally relied on the forests and wildlife of Khao Yai National Park to tide them over during times of food shortage. Without alternative livelihoods, they inevitably will deplete the park's resources.

- The Everglades, though it is an enormous national park, covers less than one-tenth of a much larger system of wetlands in southern Florida. Anything that disrupts water flow in that larger area—drainage, diversion of water for agriculture, construction of dikes and canals—will change the character of the ecosystem contained within the park.

- The animal species protected in Amboseli National Park in Kenya are for the most part not sedentary, but migratory. During the rainy season, they naturally range far beyond park boundaries, where they encounter farmers, poachers, and other competitors for their habitat. When elephants changed their behavior and ceased migrating because Amboseli offered a refuge safe from poaching, they gradually degraded the ecosystem and made it less capable of supporting other species.

Parks are not sanctuaries. They are beset by an array of forces that influence their ability to sustain biodiversity. In addition, even the best protected parks may be simply too small for the needs of some species. Parks are literally remnants of what were once vast, continuous ecosystems. Recall from "Only One Atmosphere" (Unit 4) how global warming by the greenhouse effect may affect the Everglades. Parks are not oases, but pieces of a planet undergoing profound change. How do you think climate change might affect some of the other ecosystems shown in "Remnants of Eden"?

Balancing Conservation with Local Needs

For a long time, it was believed possible to establish parks and police their boundaries, keeping local people out. Conservationists now recognize that this is unrealistic and that biodiversity cannot be protected if local people receive no benefit from its protection. Reconciling parks and local needs is rarely easy, but experiences in several parts of the world demonstrate that it is at least possible.

- In Thailand, the Population and Community Development Association (PDA) and the Wildlife Fund of Thailand have joined forces to assist communities like Ban Sap Tai with livelihoods that don't depend on resources harvested illegally from Khao Yai National Park. Their plan involves both education and economics—new kinds of farming, new forms of credit, new access to family planning and health services.

- In Kenya, the Maasai tribespeople whose farms and livelihoods were formerly threatened by migrating wildlife now have a stake in that

wildlife. By setting up small tourist camps outside the Amboseli National Park, the Maasai benefit from income brought by wildlife-viewing visitors and help maintain the annual migrations on which Amboseli's wild herds depend.

- In Costa Rica, Daniel Janzen considers parks to be "biocultural reserves" that provide an important educational and cultural resource for the people. Thinking of parks this way requires political support too, as explained by the country's Minister of Natural Resources, Alvaro Umana. If leaders sense that ecological integrity and biodiversity are important to their citizens, they will set priorities and devote resources to those goals. Costa Rica has gone farther than most nations in making protection of biodiversity a national goal.

Umana says that Costa Rica has been able to achieve so much in part because it abolished its army 40 years ago and has been able to devote resources to conservation and education that other nations have used to maintain military forces. Is this a choice that you think other countries should consider?

Undoing the Damage

"Remnants of Eden" shows a number of examples of attempts to reverse human damage to ecosystems—and also hints at the limits of these restoration activities:

- The reintroduction of the Arabian white oryx at the Yalooni Research Station in Oman shows the expense and difficulty of restoring even one species to its native habitat—even when its habitat is essentially intact. The reintroduction appears to be an initial success, but it is impossible to now know the long-term implications. The oryx population was reduced to a tiny "world herd" of only nine individuals kept in zoos, and when a population shrinks to this size, its descendents remain genetically vulnerable to extinction for many generations. The dynamics of small populations and how they predispose some species to extinction are only now being unraveled by the new science of "conservation biology."

- Restoration of whole ecosystems can be both easier *and* much more difficult than single-species reintroductions. It is easier because in many cases changing one physical factor—the flow of water, the presence or absence of fire—can change the entire character of an ecosystem and the species it sustains, providing there is a source for native species to recolonize the site. It is more difficult because changing such determinants of diversity often involves managing or restraining powerful social and economic forces.

- Both Guanacaste National Park in Costa Rica and the Everglades in Florida demonstrate that representative diversity can in some cases be restored to complex ecosystems. Changes in the species composition and entire trophic structure can result. But full restoration takes time—in some cases, centuries or longer.

- Full restoration also assumes the presence of all the constituent species. Extinct species cannot be restored. And often it is the "little things that run the world." Invertebrates and other inconspicuous species, whose disappearance might go unnoticed, often play ecological roles crucial to the functioning and resilience of an entire ecosystem.

Take a Closer Look at the Featured Countries

Costa Rica, a Central American country no larger than West Virginia, contains an array of biological diversity far disproportionate to its size (Figure 7-1). With volcanic mountains, coastal plains, and a tropical climate, the country sustains many types of ecosystems including rich rain forests and cloud forests. These varied habitat types support a breathtaking assortment of flora and fauna that includes 848 species of birds, 205 mammals, 218 reptiles, and over 9,000 species of plants—about 4% of all plant species on earth.

Compared to its neighbors in the region, Costa Rica is sparsely populated with about 3.6 million people, but its population growth is high at nearly 1.8% each year. Although Costa Rica is by no means free of poverty, it has achieved a standard of

Figure 7-1   Costa Rica

living unusual among developing countries: Literacy is nearly universal, infant and child deaths are the lowest in Latin America, and life expectancy, at 77 years, is comparable with that in the United States. Most people work in agriculture, with coffee, beef, and tropical fruits making up a large share of the country's exports.

Much of the stability and relative prosperity in Costa Rica can be attributed to a tradition of democratic government dating to the country's independence from Spain in 1848. In addition, the country abolished its army in 1949 and has made social welfare its top priority. In 1987, President Oscar Arias was awarded a Nobel Peace Prize for his efforts to pursue peace through negotiation in Central America.

This stability has allowed the country to make an unusual commitment to conservation: More than 11% of its land is designated as national parks and reserves. Including state lands on which timber harvests and other uses are allowed, the share of Costa Rica devoted to conservation uses is fully one-quarter—the highest in the world. The parks are exceptionally diverse, ranging from pristine Pacific beaches to steamy lowland rain forests to mist-shrouded cloud forests on the volcanic ridges of the continental divide.

Yet its biological diversity is not entirely secure. Outside its parks and reserves, Costa Rica has one of the highest rates of deforestation in Latin America, driven by small farmers seeking new fields and pastures. And like many of its neighbors, the country is burdened by foreign debt, which it struggles to repay. Economic hardship has forced the country's leaders to cut back on social welfare programs, and Costa Rica has sought new ways to meet its debt obligations without increasing the burden on its people. Costa Rica was among the first countries to enact "debt-for-nature" swaps, in which agreements to protect valued ecosystems are exchanged for partial debt relief.

Kenya, an East African nation of 29 million people, was formerly a British colony that gained independence in 1963 (Figure 7-2). Located on the equator, the country has a semiarid tropical climate and a landscape dominated by savanna grasslands. Most well-known is the Serengeti Plain, shared with Tanzania to the south, on which can be found the most diverse assemblage of large mammals remaining on earth. The Serengeti ecosystem, with its herds of antelopes and wildebeests, elephants, zebras, giraffes, lions, and cheetahs, has attracted scientists and tourists from all over the world. It is one of the last places where one may witness a world that our prehistoric ancestors might have known.

Like many newly independent African nations, Kenya is a poor country struggling to develop and to improve living conditions. The obstacles are daunting. Kenya's population growth, about 2.1% each year, is one of the world's highest. Some 46% of the population is under 15 years old. Infant mortality is high, illiteracy is widespread, and life expectancy is 49 years. The gross national product per person is only $340. Most Kenyans live in rural areas and farm for subsistence; the country's main exports are coffee and tea, and its main food crop is maize (corn).

Kenya's unique wildlife is both a luxury and a necessity for the country. With an exploding population, the country needs farmland—and the wildlife areas are viewed as a safety valve for the country's future food needs. Farmers already invade national parks and game reserves in search of new cropland and grazing for livestock. But the wildlife also brings tourists, and tourism earns Kenya more foreign exchange than does any export. To maintain these earnings, which it needs to buy oil, manufactured goods, and other essentials, Kenya has to reconcile the protection of its wildlife with the needs of its people. The choice is by no means painless.

Kenya has struggled to maintain its biological wealth, and keep attracting tourists, without overlooking its people's needs for land, food, and

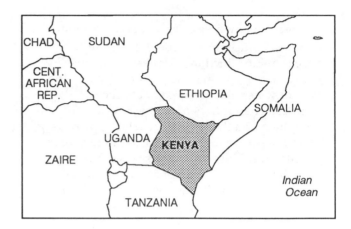

Figure 7-2   Kenya

jobs. National parks, created while Kenya was still a British colony, have been maintained since independence; the country's 29 protected areas cover more than 5% of Kenyan territory. Yet outside those parks and reserves, Kenya has lost an estimated 48% of its wildlife habitat to farmland and settlement.

While the country's treasury has continued to benefit from the foreign exchange brought by tourists, only in recent years has the need to include local people *directly* in benefits from wildlife, as shown in "Remnants of Eden," become clear. Illegal poaching of elephants and rhinos pushed both of those species into a catastrophic decline despite the country's parks and game reserves, and Kenya became the first African nation to endorse a worldwide ban on ivory trade in the summer of 1989, a last-ditch effort to save the elephant. But even if it can be protected from poaching, Kenya's wildlife is unlikely to survive to benefit the country and inspire people around the world if the country's population continues to double every 20 years.

The Sultanate of **Oman** lies on the Arabian Peninsula, hugging the coastal plain to the east of Saudi Arabia (Figure 7-3). Its sparse population of 2.5 million is growing by 3.9% each year, rapidly enough to double in 18 years. An Islamic country in the oil-rich Middle East, Oman is considerably more well-off in financial terms than most Third World countries but still suffers from widespread illiteracy and high infant and child mortality. Oman is an oil producing and oil exporting country but lacks the vast oil reserves of its neighbors on the Persian Gulf. It has remained rather isolated

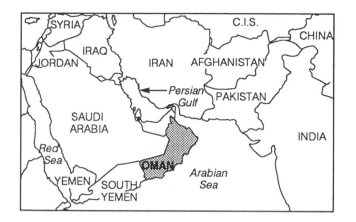

Figure 7-3   Oman

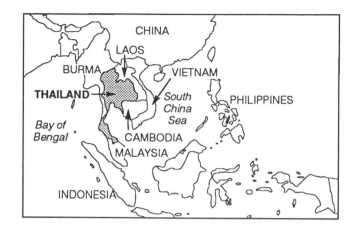

Figure 7-4   Thailand

from the world economy and the effects of large-scale oil drilling and transport.

Today Oman has a desert climate, harsh and forbidding. Such a climate does not support richly varied flora and fauna, but the plants and animals that live in deserts possess remarkable adaptations to the hot, dry conditions. The white oryx shown in "Remnants of Eden" is an example, getting nearly all of its water from the plants it eats and conserving it so effectively that it can go nearly a year without drinking. Though Oman does not have great numbers of species, these special adaptations and the austere beauty of the desert make the country's biodiversity worth maintaining.

Oman has only two protected areas, covering less than 1% of the country's territory. Successful conservation of Oman's species and habitats is thus likely to depend on involvement of the country's nomadic Bedouin peoples, like the Harasis tribesmen shown in the program who range with their livestock over much of the country.

The Kingdom of **Thailand**, on the mainland of Southeast Asia between Burma, Laos, and Cambodia, lies entirely within the tropics (Figure 7-4). Although it is still primarily an agricultural country, Thailand in recent years has become known as one of the "newly industrializing countries" whose rapid economic growth is due to electronics, textiles, and products manufactured for export. But export crops—for example, rubber and palm oil and tropical hardwoods cut from the country's forests—have traditionally been the economy's backbone. Thailand's biological diversity has been altered by pervasive agricultural settlement, logging, and, increasingly, by the

consequences of urbanization and affluence.

The country's population of 62 million is growing by 1.1% each year, one of the lowest rates in Southeast Asia but still enough to double the population in 64 years. Thailand's imaginative and well-publicized family-planning program (profiled in Unit 11) has slowed the birth rate dramatically, helped by the country's strong economy. Continued progress in stabilizing the population will relieve pressure to clear the country's last intact forests for farmland.

Located in the humid tropics, Thailand was originally more than 70% forested. Abundant rainfall throughout the country and the mountainous terrain of the north favor a great profusion of plant and animal species. But villagers' need for rice paddies and a seemingly endless supply of high-value timber trees led to rapid deforestation; today, forests stand on less than 20% of Thailand. The country learned that those forests, particularly in the hilly north, protected more than rare species. Heavy downpours in late 1988 rushed off deforested slopes, flooding hundreds of villages and killing scores of people. The tragedy and subsequent outcry led Thailand's Prime Minister to make an unprecedented decision: Ban commercial logging and revoke farming rights in remaining forests. Though unlikely to be permanent, the logging ban gave Thailand a chance for a long-overdue reappraisal of its forest-use policies.

Khao Yai National Park, shown in "Remnants of Eden," was the first national park in the country; today, 65 parks and protected areas encompass 7.8% of Thai territory. With effective protection of these parks, attention to the economic needs of settlers and communities on park boundaries, and sustainable management of the country's remaining forests, it may be possible to keep plant and animal extinctions from accelerating.

The **United States**, though not a tropical country, is large and varied enough to support an enormous spectrum of ecosystems ranging from extreme desert to Arctic tundra (Figure 7-5). These ecosystems support a rich native diversity of plant and animal species. The continental United States has an estimated 22,000 native plant species and 1,181 vertebrates (mammals, birds, reptiles, amphibians, fish). But to put this temperate-zone biodiversity into perspective, tropical Central America, comprising only one-twentieth as much territory as the United States, has a comparable number of plant species and twice as many native vertebrates.

The country that pioneered the national parks concept (Yellowstone National Park was established in 1872) now has 370 units, including 54 major parks, numerous wildlife reserves, and other protected areas covering 7.4% of national territory. The Everglades, featured in the program, is one of 50 national parks and is considered perhaps the most endangered in the system. The causes of that endangerment are due primarily to the multiple human demands in the region on the water—the lifeblood of the Everglades ecosystem. The example demonstrates that reconciling biodiversity with human needs is not easy even in an affluent society.

The United States has done more than most countries to protect and manage its biological diversity—by creating parks and other protected areas, by passing laws like the Endangered Species Act, and by supporting research on ecosystems and species. But a national inventory of biological diversity has never been carried out, research has shown that nearly all the national parks are losing animal species, and remaining wilderness areas are beset by pressures ranging from wetland drainage to logging to air pollution and acid rain. The prospect of climate change, which will be especially pronounced in temperate-zone countries such as the United States because of temperature shifts, holds out the likelihood of ecological changes of unknown magnitude. The 273 million residents of the wealthiest industrial society face a biological future as uncertain as any nation on earth, and

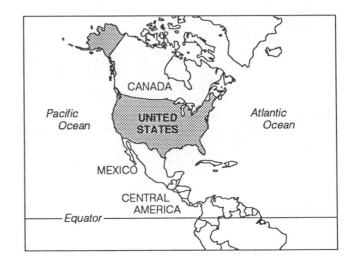

Figure 7-5   The United States

much work in understanding biodiversity and learning how to coexist with it remains to be done at home.

Examine Your Views and Values

1. Do you agree with Edward O. Wilson of Harvard University that human activities which destroy wild habitats and hasten the extinction of species are equivalent to "burning Renaissance paintings to cook dinner"? Why or why not?

2. Would you be as supportive of efforts to protect endangered plant and insect species as of those to protect large, familiar species such as pandas or elephants? Why or why not?

3. Most religions include explicit or implicit teachings about human responsibility toward other creatures. If you are a member of a particular faith, what does that faith teach about human obligations toward other species or toward the natural world? If you attend religious services, discuss the question with a member of the clergy. Are there religious reasons for concern about the widespread extinction of plant and animal species?

TEST YOUR COMPREHENSION

Self-Test Questions
(Answers at end of Study Guide)

Multiple Choice

1. The Arabian white oryx was pushed to the brink of extinction because
   a. its forest habitat was converted to desert.
   b. it was overhunted even though its habitat remained intact.
   c. the Harasis Bedouin overused the oryx for food and hides.
   d. all species become extinct sooner or later.

2. Released oryxes need to be monitored closely because
   a. they might eat the wrong foods in their unfamiliar surroundings.
   b. they need protection from hunters and poachers.
   c. their breeding must be managed to reduce in-breeding.
   d. they could damage the fragile desert ecosystem.

3. According to biologist Terry Erwin, tropical forests have many insect species because
   a. the complicated structure of the forest gives them many places to live.
   b. such forests grow in areas that are much more fertile than are other ecosystems.
   c. many species have moved there from other ecosystems.
   d. there are fewer birds and other insect-eating species than elsewhere.

4. The great hornbill was chosen for study in Thailand because
   a. it is found in many parts of the country and is easily studied.
   b. it is able to survive drastic modifications of its habitat.
   c. it is used for food by the villagers of Ban Sap Tai.
   d. it depends on undisturbed rain forest for its survival.

5. The villagers of Ban Sap Tai hunted animals and cut wood in Khao Yai National Park because
   a. they needed those resources in years that their crops failed.
   b. they were displaced from their former lands when the park was created.
   c. they had rights under an agreement with the government.
   d. they considered the park's resources to be unlimited.

6. The purpose of the Rural Development and Conservation Plan in Ban Sap Tai was
   a. to give villagers alternative livelihoods that reduced their need to invade the park.
   b. to give authorities more power to punish villagers who damaged the park.
   c. to help villagers make more money from the park's wildlife and timber.
   d. to turn the park over to control by local people.

7. The research in Lake Naivasha is intended to show that
   a. a lake ecosystem can be protected from changes in the surrounding environment.
   b. the lake was not productive until American fish species were introduced.
   c. the lake food chain can be manipulated to support a bigger catch.
   d. the fish catch can be expanded indefinitely to meet the population's needs.
   e. the lake ecosystem should not be modified by human beings.

8. The biodiversity of the Everglades ecosystem
   a. supplies an important food source needed by the people of Florida.
   b. depends on a water source that is also diverted for human uses.
   c. is protected from urban and agricultural development by its national park status.
   d. has escaped serious damage from human activities so far.

9. Restoring the flow of the Kissimmee River
    a. has been unable to reverse the loss of biodiversity in the Everglades.
    b. has brought back a tree species that allows other species to recover.
    c. has brought back a tree species whose leaves can't be digested by other animals.
    d. has re-created the Everglades as it was a century ago.
    e. has made fires far more common than previously in the Everglades.

10. Daniel Janzen compares the impact of fire in Guanacaste National Park to that of an exotic animal species because fire
    a. consumes the basic production of the ecosystem just as a large animal might.
    b. was unknown in Costa Rica and was brought there by settlers.
    c. makes the park's ecosystem more productive.
    d. is a natural element of Costa Rica's biological diversity.

11. The Costa Rican government official interviewed in "Remnants of Eden" believes that
    a. environmental protection is a luxury that poor countries cannot afford.
    b. environmental protection is more important than poverty, education, and health.
    c. reducing military spending is one source of money needed for conservation.
    d. politicians should not be involved in conservation matters.

Sample Essay Questions

1. Define indicator species and cite an example from the television program to explain how scientists use such species to gauge the health of ecosystems.

2. Explain why the television program "Remnants of Eden" emphasizes the ecological importance of insects and other invertebrate species.

3. Choose one ecosystem shown in "Remnants of Eden," describe its characteristic diversity, and explain how it demonstrates the basic ecological concepts of energy flow and nutrient cycling.

4. Compare the impacts of the Maasai tribespeople in Kenya and the villagers of Ban Sap Tai village in Thailand on ecosystems and biological diversity. How have they been included in current efforts to protect biodiversity?

5. Describe and contrast two efforts to *restore* ecosystems.

GET INVOLVED

References

Baskin, Yvonne. *The Work of Nature: How the Diversity of Life Sustains Us.* Washington, D.C., 1997.

Durning, Alan T. *Saving the Forests: What Will It Take?* Washington, D.C.: Worldwatch Institute, 1992.

Ehrlich, Paul R. *The Machinery of Nature.* New York: Simon & Schuster, 1986.

Leopold, Aldo. *A Sand County Almanac.* New York: Oxford University Press, 1949.

Myers, Norman. *A Wealth of Wild Species: Storehouse for Human Welfare.* Boulder, CO: Westview, 1994.

National Academy of Sciences. *Biodiversity II. Understanding and Protecting Our Biological Resources.* Washington, D.C.: National Academy Press, 1996.

Pimentel, David, et al. "Economic and Environmental Benefits of Biodiversity." *BioSciences,* December, 747-57.

Ryan, John C. *Life Support: Conserving Biological Diversity.* Washington, D.C.: Worldwatch Institute, 1992.

Terborgh, John. *Requiem for Nature.* Washington, D.C.: Island Press, 1999.

Wilson, E. O. *The Diversity of Life*. Cambridge, MA:
Harvard University Press, 1992.

Wilson, E. O. "Threats to Biodiversity." *Scientific
American*, Sept. 1989.

World Commission on Environment and
Development. "Species and Ecosystems: Resources
for Development." *Our Common Future*. Oxford:
Oxford University Press, 1987.

Two chapters in *Taking Sides: Clashing Views on
Controversial Environmental Issues* consider the value
of wilderness and the legal protection of
endangered species in the United States:

- Issue 1. Does Wilderness Have Intrinsic Value?
  (pp. 2–23)

- Issue 2. Do We Need More Stringently Enforced
  Regulations to Protect Endangered Species?
  (pp. 24–39)

Also see chapters on biodiversity in various
editions of Worldwatch Institute's annual *State of
the World* report:

- Chris Bright, "Anticipating Environmental
  Surprise," *State of the World 2000*.

- Janet N. Abramovitz and Ashley T. Mattoon,
  "Reinventing the Forest Products Economy,"
  State of the World 1999.

- John Tuxill, "Appreciating the Benefits of Plant
  Biodiversity," *State of the World 1999*.

- Janet N. Abramovitz, "Sustaining the World's
  Forests," *State of the World 1998*.

- John Tuxill and Chris Bright, "Losing Strands in
  the Web of Life," *State of the World 1998*.

- Sandra Postel, "Reforming Forestry," *State of the
  World 1991*.

- Edward C. Wolf, "Avoiding a Mass Extinction
  of Species," *State of the World 1988*.

- Edward C. Wolf, "Conserving Biological
  Diversity," *State of the World 1985*.

Organizations

Many local groups and national organizations are
actively involved in studying biodiversity and
taking steps to protect it. Local chapters of national
environmental groups such as the Sierra Club and
the National Audubon Society are good sources for
information about nearby ecosystems, as well as
national and international problems. Zoos, aquar-
iums, natural history museums, and botanical
gardens sponsor speakers and other public
programs concerned with biodiversity issues.

If you want to learn more about what is being
done to protect species and ecosystems worldwide,
particularly in tropical countries, contact any of the
following organizations:

Conservation International
2501 M. St. NW, Suite 200
Washington, DC 20037
Tel: (202) 887-0193; web: www.conservation.org

National Wildlife Federation
8925 Leesburg Ave.
Vienna, VA 22184
Tel: (703) 790-4000; web: www.nwf.org

The Nature Conservancy
4245 North Fairfax Drive, Suite 100
Arlington, VA 22203-1606
Tel: (703) 841-5300; web: www.tnc.org

World Wildlife Fund—U.S.
1250 24th Street NW
Washington, DC 20037
Tel: (202) 332-2200; web: www.worldwildlife.org

UNIT 8

More for Less

The development of wind farming as a renewable source of power has been pioneered in California, Denmark, India, and the Netherlands.

BEFORE YOU VIEW THE TELEVISION PROGRAM

Learning Objectives

After completing the assigned readings and viewing "More for Less," you should be able to

- explain the difference between *renewable* energy resources and *nonrenewable* energy resources, and cite examples of each.

- describe the principal environmental impacts associated with the production and use of crude oil, coal, and natural gas.

- list the environmental advantages and disadvantages associated with nuclear power.

- identify three renewable energy sources, describe their current contribution to energy supplies, and evaluate their potential to meet future energy needs.

- describe how you would go about exploring opportunities to use energy more efficiently in your household, workplace, university, or community—and explain why.

- define net useful energy, as it relates to solar energy, fossil fuels, and nuclear power.

Reading Assignment

Choose the material from either textbook as your reading assignment. Your instructor might assign additional readings as well.

Living in the Environment

Chapter 3, "Matter and Energy Resources: Types and Concepts," to review basic energy concepts if necessary. Read the Osage, Iowa story on pp. 56-57.

Chapter 15, "Nonrenewable Energy Resources"

Chapter 16, "Energy Efficiency and Renewable Energy Resources," entire chapter, and Amory Lovin's Guest Essay on pp. 433-434

Environmental Science

Chapter 3, "Science, Systems, Matter, and Energy," to review basic energy concepts if necessary

Chapter 19, "Nonrenewable Energy Resources"

Chapter 20, "Energy Efficiency and Renewable Energy"

Unit Overview

This unit of *Race to Save the Planet* surveys the human use of energy resources, considers the environmental problems caused by energy use in industrial and developing societies, and introduces innovative ways of increasing energy supply and managing energy demand that can help us cope with these problems.

The assigned text chapters introduce **nonrenewable** energy resources (fossil fuels, geothermal energy, and nuclear energy) and **renewable** energy resources (conservation, solar energy, wind, water, and biomass). The magnitude of energy supplies, technologies for energy conversion, controversies over the use and environmental impacts of different technologies, and design of appropriate national energy strategies are discussed in detail.

The accompanying television program, "More for Less," documents the search for energy solutions. Individuals in both industrial and developing countries are pioneering energy sources that are affordable, politically acceptable, and less damaging to the environment than are the conventional fossil fuels that form the backbone of the world's present energy system. The program explores ways to reduce energy demand by increasing energy efficiency; highlights efforts to redesign conventional coal-burning, wood-burning, and nuclear technologies into more sustainable "transition technologies"; and visits communities around the world that have begun to make the transition to renewable power sources.   Brazil, Denmark, India, and the United States are featured in the program.

This unit is intended to equip you to make more informed choices about energy—with an appreciation of the complexity of the world's present energy system and a thorough grasp of the alternatives.   Whatever course the energy transition takes, it is going to change our world.

Glossary of Key Terms and Concepts

The following terms and concepts will be helpful as background for viewing "More for Less."

Bagasse is a solid material left after sugar cane is crushed and its pulp extracted for fermentation into alcohol.   Brazil discovered in the course of its national program to develop alcohol fuels that bagasse, formerly discarded as waste, could be burned economically to generate electricity and also processed into a feed for livestock.

Biogas is a combustible gas (composed primarily of methane) produced when sewage or manure is fermented in the absence of oxygen.   Manure-based biogas digesters that draw their raw material from the country's dairy farms are used as part of Denmark's LOCUS plants (see below).   In India, China, and other parts of Asia, household-scale biogas digesters are used to produce a cooking fuel from human and animal wastes.   The solid material that remains in the digester after fermentation can be used as an organic fertilizer.

Cogeneration refers to a facility in which two or more forms of energy are generated simultaneously or interchangeably.   Commonly, a cogeneration facility produces steam for an industrial or commercial process and uses some of the steam to turn a turbine that generates electricity.   Another type of cogeneration arrangement combines several energy sources in a single facility to provide a mix of energy forms (heat, electricity, etc.) in varying proportions according to the needs of energy users.

Based on the same principle as the fluorescent tubes that light businesses and offices, each **compact fluorescent light bulb** supplies as much light as a conventional incandescent light bulb but requires only a fraction of the electricity.   Using 11 watts of power, a compact fluorescent bulb can be screwed into normal sockets, provide as much light, and last six times as long as the standard 60-watt incandescent bulb it replaces.

A **district heating plant** is a central boiler from which steam is distributed for heating to homes in the surrounding community.   This type of facility distributes an energy service—heat—rather than an energy source to its customers.

The **high-temperature gas-cooled reactor** is a form of nuclear fission reactor in which the fuel elements are cooled by circulating helium gas.   The different fuel configuration and comparatively small size (generally about one-tenth the size of a standard 1000-megawatt power plant) of this type of reactor supposedly make it invulnerable to loss-of-coolant accidents (see below).

The **LCP 2000** ("Light Component Project") is a prototype high-efficiency automobile designed by Volvo to minimize the energy required when it is manufactured and driven.   Constructed with lightweight materials, the LCP 2000 weighs half as much as an average American car and is powered by an advanced diesel engine that can run on a variety of fuels.   Its estimated fuel consumption is 98 miles per gallon.   Volvo has no current plans to market the LCP 2000.

Denmark pioneered a cogenerating technology known as **LOCUS** (for "Local Cogeneration Utility System") that combines several different renewable energy sources so that it can supply a mix of energy services to consumers as needed.   A typical facility combines wind-generated electricity and a boiler fired by biogas that can supply steam for heating.   LOCUS plants are community-scale and designed to capture locally available flow-energy sources.

A **loss-of-coolant accident** occurs when the circulation of coolant through the fuel assemblies of a nuclear fission reactor is halted or interrupted, allowing the heat generated by fission reactions in the reactor core to build up to the point that it threatens to melt the fuel elements and the reactor itself. If the reactor's containment vessel is damaged, large amounts of radioactivity may be released into the environment. The most serious accident in the history of the U.S. nuclear power industry, at Three Mile Island in Pennsylvania in 1979, was a loss-of-coolant accident.

Energy analyst Amory Lovins coined the term **"negawatts"** to signify that each unit of energy saved amounts to exactly the same thing as an additional unit of energy supplied. As long as the steps to save energy are cheaper than the cost of generating an equivalent amount of additional energy, a utility or business has more incentive to invest in negawatts than in megawatts.

The two **oil crises** of the 1970s (in 1973 and 1979) were caused by restrictions on oil exports from parts of the Middle East, which curtailed world oil supplies enough to cause a sudden, abrupt increase in the price of crude oil. Although the crises were caused by political, rather than physical, scarcity of oil, they revealed the risks of the world's heavy reliance on the oil reserves of the Middle East and stimulated a search for alternatives to continued reliance on fossil fuels.

Photovoltaic cells convert the sun's energy directly into electric current. When sunlight strikes certain materials, it dislodges electrons, which can be channeled to flow as current. Researchers in Japan, Europe, and the United States have steadily improved the efficiency and reduced the cost of photovoltaic cells; they are soon expected to compete with other large-scale sources of power generation. Photovoltaic cells are already widely used to supply electricity at remote installations, on satellites, and to operate calculators and other small electronic devices.

Pressurized fluidized-bed coal combustion is an alternative technology in which pulverized coal is injected into a furnace onto a bed of heated limestone, agitated by forced air. Combustion takes place at half the temperature of conventional furnaces, producing 50% fewer nitrogen oxides, and the sulfur released as coal burns reacts with the limestone rather than escaping as sulfur dioxide. The process is the most efficient way known to produce electricity from coal. Although they control the two most serious pollutants associated with coal burning, fluidized-bed boilers do emit carbon dioxide, adding to the greenhouse effect.

A **utility** supplies a service—typically water, electricity, gas, etc.—to the general public. Some utilities, for example most water systems, are government-run, whereas others, such as telephone and power companies, generally operate in the private sector in the United States. Electric utilities manage the generation and distribution of electricity, deciding when new power plants need to be built and how much customers should be charged. Because they have a virtual monopoly on the service they provide, utilities are overseen by regulatory commissions that determine which costs can be passed along to customers. The degree of public participation and responsiveness shown by the Osage Municipal Utility, featured in the television program, is unusual in the utility industry.

AFTER YOU VIEW THE TELEVISION PROGRAM

Consider What You Have Seen

"More for Less" documents the search for energy solutions—the first tentative steps away from a fossil-fuel–based energy system. The case studies and examples presented in the program illustrate three principal (and interrelated) themes in the effort to reduce the environmental toll of the present energy system while supplying needed energy services:

- Putting energy efficiency to work

- Designing transition technologies that reduce environmental damage

- Shifting to renewable power sources

Putting Energy Efficiency to Work

Four of the program's case studies deal with harnessing energy efficiency:

- Osage, Iowa's community-wide energy - efficiency program

- Physicist Amory Lovins's presentation of the advantages of "negawatt technologies"

- Automotive designer Rolf Mellde and Volvo's LCP 2000

- Amulya Reddy and fuel-efficient wood stoves in India

As you reflect on these sequences, keep the following themes and concepts in mind and review the reading assignment if necessary.

Renewable vs. nonrenewable energy resources: What energy sources are at issue in each sequence? What makes them either renewable or nonrenewable?

The role of energy efficiency: What energy services (e.g., lighting, heat, energy for cooking, steam for industrial use) are the people in each sequence concerned with, and what options exist in providing these services? Are any options overlooked in each case?

Economic factors: How is the price of energy involved in this sequence? How do changes in energy prices affect efforts to use energy more efficiently?

The decision-making process: In each sequence, who is in a position to decide how efficiently energy is used? What influences their decisions to pursue efficiency?

Industrial and developing societies: Based on the cases shown, does the rationale to use energy more efficiently differ in industrial and developing societies?

Designing Transition Technologies

Three sequences deal with designing transition technologies that can reduce the environmental damage associated with conventional energy technologies today:

- Fuel-efficient wood stoves in India

- Pressurized fluidized-bed coal combustion

- Testing an "inherently safe" nuclear reactor

As you reflect on these sequences, keep the following themes and concepts in mind and review the assigned readings if necessary.

Environmental consequences: List the environmental concerns associated with producing and using the fuel source that each of the transition technologies featured depends on. What environmental damages or risks are reduced by the technologies shown? What environmental concerns are not addressed?

Energy prices: How is the adoption of each technology featured likely to be affected by changes in the price of the source (e.g., wood, coal, uranium) whose energy it is designed to capture? How do you think it would be affected by changes in the price of alternatives?

Energy services: What form of energy is supplied by each technology shown? What sorts of needs are met by energy in this form?

Decision making: Who is in a position to decide whether the technology in question is adopted? Where does this technology fit into the overall energy system?

Global perspective: How much of the world's present energy use is supplied by the *source* each featured technology is designed to use; for example, how much of the world's present energy supply is based on wood, coal, nuclear power?

The geography of energy: In what regions of the world do you think each featured technology is likely to make its largest contribution? Is each more relevant to the energy needs of industrial or developing societies?

Shifting to Renewable Power Sources

Three sequences in the program present communities that have begun to shift toward reliance on renewable energy sources to meet their energy needs:

- Brazil: the national alcohol fuels program

- Denmark: Bornholm Island's community energy plan

- India: photovoltaic cells

As you reflect on these sequences, keep the following themes and concepts in mind and review the assigned readings if necessary.

Energy resources: What energy sources are being considered in each instance? What limitations and constraints are they subject to because they are renewable resources? How do these constraints influence the way each community plans to use each source in meeting its energy needs?

Economic factors: What concerns about the cost of energy prompted a consideration of renewable energy sources in each case? What are the costs of relying on these power sources?

Energy geography: How is the contribution of renewable energy sources likely to differ in the industrial and developing societies featured in the program? What sources seem best suited to the needs of societies at different economic levels?

Decision making: Who is in a position to decide when and how renewable energy sources are adopted in the three featured cases? What concerns influence their decisions?

The transition to a sustainable energy system: What conventional energy sources would be displaced by the renewable sources adopted in each case? What broader environmental and economic problems associated with energy use would this substitution address?

Take a Closer Look at the Featured Countries

Although **Brazil's** 168 million citizens live in what is considered a newly industrialized society, more than a third of the energy used in the country still comes from noncommercial sources, primarily the firewood needed for heat and cooking in poor rural households (Figure 8-1). But considering the size of its cities and manufacturing centers, Brazil already depends to an unusual degree on domestic renewable energy sources.

The country has tapped many of the tributaries of the Amazon basin for hydroelectricity (in some cases with huge projects that have increased Brazil's foreign debt) to supply power to industries and urban areas, and charcoal is widely used for smelting in the steel industry. Alcohol distilled from sugar cane has become Brazil's primary automotive fuel, helping to reduce the country's dependence on imported oil. The bagasse waste left after alcohol production is now burned to generate electricity. Despite progress on renewable fuels, Brazil still imports a significant amount of oil each year.

The country once had ambitious plans for a national nuclear power program, but Brazil's one operating nuclear reactor was closed after the Chernobyl accident in the former Soviet Union in 1986, and nuclear energy is not expected to play a significant role in the future.

Having made a substantial head start on a nationwide transition to reliance on renewable energy sources available within its borders, Brazil still faces considerable hurdles. The hydroelectricity program has been costly, and proposed dams have sparked opposition on social and environmental grounds. Although preferable to importing oil, the alcohol program has been a drain on the national treasury and has increased competition for fertile land, fueling the exodus of rural families displaced from farmland into the forests of the Amazon basin. And many industrial projects depend on charcoal made from virgin forest rather than tree plantations, accelerating the deforestation of the Amazon region.

Figure 8-1   Brazil

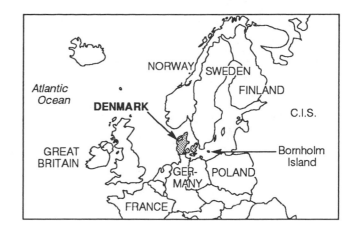

Figure 8-2   Denmark

With a population expanding by nearly 3 million each year, Brazil needs to expand its capacity to supply energy services. Much of the country's present energy use is needlessly wasteful, and opportunities to increase energy efficiency offer the most affordable means of supplying those needed services. According to one study, $10 billion invested in more efficient refrigerators, street lights, and improved industrial motors could eliminate the need to invest $44 billion in new power plants and dams to generate electricity for urban and industry markets. If Brazil really intends over the long run to live within the bounds of its renewable resources, it will have to restrain its population growth as well.

Like its northern European neighbors, **Denmark** ranks among the most energy-efficient industrial societies on earth, using half as much energy as the United States for each unit of economic output and just over half as much energy per person (Figure 8-2). A small nation (5.3 million people) that has almost attained zero population growth, Denmark is spared the problem of having to manage a rapidly expanding demand for energy. However, the country has few fossil fuel resources of its own and depends heavily on imported fuels, especially petroleum. With 1.5 million cars (one for every 3.5 citizens), Denmark has a considerable demand for liquid fuels that cannot easily be satisfied with domestic resources. Although the country has no hydroelectric generating potential to speak of, Denmark's Parliament voted in 1985

never to develop nuclear power and even asked neighboring Sweden to close an existing reactor at Barsebäck.

Given Danish attitudes and the country's sparse endowment of fossil fuels, it is not surprising that Denmark has been a pioneer in developing renewable energy sources. Denmark is a world leader in wind power technology and exports its wind turbines to many other countries. It has 100 megawatts of installed wind-generating capacity at home and intends to double that figure by 1991. Plans to develop additional renewable power sources focus on the use of straw, manure, and other agricultural wastes.

Denmark has been in the forefront of efforts to develop energy-efficient technologies such as low-wattage compact fluorescent light bulbs and innovative power cogeneration arrangements like the LOCUS plants, which combine wind-generated electricity and steam heat produced by manure-derived biogas. Gasoline use is discouraged by Denmark's high fuel taxes; in 1987, a gallon of gasoline in Denmark cost $3.58.

With its population of 950 million growing by about 18 million each year, **India** faces energy problems of equally enormous scale and complexity (Figure 8-3). Noncommercial energy (fuelwood, crop wastes, and dung cakes burned for cooking) accounts for 54% of India's energy use, but the country's huge commercial power sector depends on fossil fuels. India produces little oil and spends approximately 80% of its export earnings on oil imports. Coal is widely used (with few environmental safeguards) in industry and by the country's railroads.

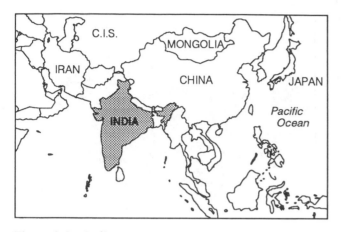

Figure 8-3   India

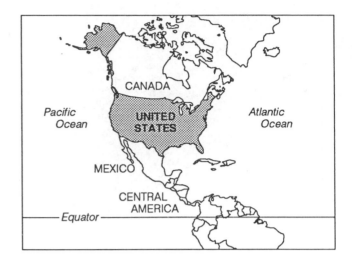

Figure 8-4   The United States

With the headwaters of its major rivers rising in the Himalayas, India has considerable hydro-electricity potential, but only 15,000 megawatts of an estimated technical potential totaling 100,000 megawatts had been installed by 1985.   In such a densely settled country, large dams inevitably displace communities and cause extensive environmental disruption.   India also had six operating nuclear reactors with a total capacity of 1,164 megawatts (about the size of one large nuclear facility in Europe or North America) by 1986, with plans for four more reactors.

India is the world's largest producer of fuelwood, but firewood shortages are endemic in parts of the country and tree cutting exceeds regrowth by a substantial margin.   More efficient wood stoves for India's poor, and a gradual shift to reliance on biomass-based liquid fuels that can be burned more efficiently, could reduce pressure on the country's woodlands.

Indian industry is among the world's least efficient, and enormous scope exists for using coal and biomass more efficiently.   Ironically, although India had only 1.4 million passenger cars in 1987 (one for every 548 people), the country has a growing domestic auto industry.   Although Indian automakers are producing models that get more than 40 miles per gallon—considerably more fuel-efficient than American cars—the wisdom of expanding automobile-based transportation in a country so populous and poor is questionable.

In addition to its basic reliance on biomass fuels, India is beginning to look to other renewable energy sources. The Ministry of Energy is promoting the installation of 5,000 megawatts of wind-powered generating capacity by the year 2000, and photovoltaic installations provide power in many villages beyond the reach of the country's power grid.   But without restraining population growth, it seems doubtful that India will ever be able to provide needed energy within the limits of the country's "energy income"—its endowment of renewable energy resources.

The **United States** stands by itself in energy production and consumption, as it does in so many other resource issues, but unfortunately the world as a whole suffers the consequences of this pro-digious energy use (Figure 8-4).   With the world's most voracious appetite for fossil fuels, U.S. energy production releases an estimated 1.2 billion tons of carbon into the atmosphere each year, more than 5 tons per person—the single largest national contribution to the greenhouse effect.   The U.S. automobile fleet alone, 150 million (more than one car for every two people), consumes as much oil as the United States imports.   The vast existing capital stock for energy generation—hydroelectric dams, nuclear and fossil-fuel–fired power plants, indus-trial boilers—and the infrastructure of wires, cables, storage tanks, and pipelines for supplying energy to users are worth trillions of dollars.

The U.S. population of 273 million is growing at 0.6% per year, an addition of 1.6 million people each year.   Because energy use per person is so high in our society, this slow population growth translates into an annual growth in demand for energy 3.6 times larger than India's, even though nearly ten children are born in India for every one born in the United States!

The energy challenge facing this country in the years ahead is not how to accommodate this growth at the present level of energy use but how to scale back the energy, particularly fossil fuels, needed to sustain living standards. The United States has made substantial progress on energy efficiency since the first oil crisis in 1973, although efficiency gains have slowed with the relatively low oil prices of the late 1980s.

Examine Your Views and Values

1. Some of the world's most damaging environmental catastrophes have occurred when conventional energy technologies fail. Examples include the explosion at the Chernobyl nuclear reactor in the former Soviet Union in April 1986 and the oil spill in Alaska's Prince William Sound when the tanker Exxon Valdez struck a reef in March 1989.

   - Do you think the primary responsibility for environmental damage in such cases rests with:
     a. the people directly responsible for the incidents?
     b. the institutions (governments and corporations) that own and operate the systems for producing and transporting energy?
     c. the societies whose energy demand justifies the existence of systems to produce energy in fragile environments and transport it long distances?

   - If accidents are inevitable in any system prone to human error, what ways would you suggest to reduce the risk that environmental catastrophes will occur?

2. List the energy *services* you depend on every day in your household. What *resources* supply those services? What environmental problems are associated with each resource on your list? How could you change your energy-use patterns to reduce your personal contribution to those problems? What considerations would affect your decision or ability to do so?

3. The Table 8-1, on the following page, shows the population and the use of commercial energy (coal, oil, natural gas, electricity) per person in the four countries shown in the television program.

Developing countries have a legitimate claim to increased energy use as their populations grow and they attempt to improve living standards. Based on what you have read and seen in the television program:

- What level of energy use per person should countries such as Brazil and India aim to achieve?

- What environmental problems would this cause or worsen?

- What responsibility do you think industrial countries like Denmark and the United States have to help countries such as Brazil and India increase their energy use in the most environmentally and economically sustainable manner?

TABLE 8-1 COMMERCIAL ENERGY USE PER PERSON IN THE 1980s

Country	1989 Population (millions)	Gigajoules Per Year[1]
Brazil	147	22
Denmark	5	156
India	835	8
United States	249	278
WORLD	5,234	56

[1] A joule is the amount of energy in a mass of one kilogram moving at a velocity of one meter per second. The prefix *giga-* means one billion.

SOURCE: Population Reference Bureau and World Resources Institute.

TEST YOUR COMPREHENSION

Self-Test Questions
(Answers at end of Study Guide)

Multiple Choice

1. One thing that was *not* part of Osage, Iowa's conservation program was
    a. infrared photographs to show homeowners where heat was leaking.
    b. insulation of water heaters and pipes.
    c. converting industrial boilers to burn straw from nearby farms.
    d. capturing and reusing heat generated by equipment at Osage businesses.

2. The compact fluorescent light bulbs demonstrated by Amory Lovins in the program show that
    a. people don't need as much light as they think they do.
    b. the same amount of light can be produced with far less energy than that needed by a conventional bulb.
    c. in contrast to other technologies, lighting accounts for only a small share of U.S. energy use.
    d. lighting offers few opportunities for using energy more efficiently.

3. The Volvo LCP 2000 has not yet been marketed because
    a. despite its lightness and efficiency, it is not as safe as existing cars.
    b. Volvo's management felt that customers were not interested in fuel efficiency.
    c. Volvo's management felt that the greenhouse effect was of no concern to auto manufacturers.
    d. most drivers cannot buy the vegetable oils the car is designed to run on.

4. Despite the low use of energy per person in India, it still makes sense to increase the efficiency of wood stoves because
    a. smoke from inefficient stoves threatens the health of Indian women.
    b. firewood shortages are common in India.
    c. more efficient stoves could be exported to industrial nations.
    d. Indian consumers want modern appliances.
    e. More than one of the above is true. The correct answers are _____.

5. Photovoltaic cells were featured in the television program because this technology
    a. is ready for widespread use.
    b. is the only known energy technology that doesn't worsen the greenhouse effect.
    c. suggests what energy sources may look like as humanity makes the transition to reliance on renewable sources.
    d. depends on sunlight, so energy efficiency is not an issue.

6. Nuclear power's contribution to slowing climate change will be limited because
    a. electricity generation is not the only use of fossil fuels that releases carbon dioxide.
    b. the Chernobyl accident in 1986 caused most countries to abandon nuclear energy.
    c. important questions of safety, waste disposal, and public acceptability have not yet been solved.
    d. nuclear power releases other greenhouse gases.
    e. More than one of the above is true. The correct answers are _____.

Fill In

1. A Danish system that combines windmills and manure-fired biogas to produce electricity and heat is called _____. This is one form of a more general arrangement called _____ in which otherwise wasted energy is tapped to provide needed energy services.

2. Brazil uses sugar cane to produce _____ and _____ as alternative energy sources. The first is a substitute for _____, whereas the second is used to generate _____.

3. The pressurized fluidized-bed technology shown in the program depends on burning _____. It reduces the emissions of _____ and _____ that cause acid rain, but like all technologies that rely on fossil fuels, it releases _____.

Matching

For each technology listed in Column A, list all of the appropriate characteristics or effects from Column B.

Column A	Column B
1. pressurized fluidized-bed coal combustion _____	a. depend(s) on *renewable* resources
2. high-temperature gas-cooled nuclear fission reactor _____	b. depend(s) on *nonrenewable* resources
3. improved wood stoves _____	c. reduced risk of melt-down
4. compact fluorescent bulb _____	d. increases carbon dioxide emissions
5. photovoltaic cells _____	e. reduces acid rain emissions
6. biogas generators _____	f. produces nuclear waste
7. windmills _____	g. improves energy efficiency
8. cogeneration plants _____	h. addresses energy needs of low-income people
9. LCP 2000 _____	

Sample Essay Questions

1. Explain this statement: "Earth's energy future will not be anything like the recent past."

2. Compare the advantages and disadvantages of the following three technologies used to generate electricity: pressurized fluidized-bed coal combustion, the high-temperature gas-cooled nuclear fission reactor, and photovoltaic cells.

3. Explain the implications of the second law of thermodynamics for the use of energy by human society.

4. How can unrestrained population or economic growth cancel out the gains achieved by societies that use energy more efficiently?

GET INVOLVED

References

Berger, John. *Charging Ahead: The Business of Renewable Energy and What it Means to America*. New York: Henry Holt, 1997.

Campbell, Colin S., and Jean H. Laberrene. "The End of Cheap Oil." *Scientific American*, March, 78-83, 1998.

Flavin, Christopher. *Sustainable Energy*. Washington, D.C.: Renew America, 1990.

Flavin, Christopher, and Nicholas Lenssen. *Power Surge: Guide to the Coming Energy Revolution*. New York: W.W. Norton, 1994.

George, Richard L., "Mining for Oil." *Scientific American*, March, 84-88, 1998.

Goldemberg, José, et al. *Energy for a Sustainable World*. Washington, D.C.: World Resources Institute, 1987.

Lovins, Amory B., and L. Hunter Lovins. "Reinventing the Wheels," *The Atlantic Monthly* January, 1995.

Lovins, Amory B., and L. Hunter Lovins. *Brittle Power: Energy Strategy for National Security*.

Andover, Mass.: Brick House, 1982.

Sant, Roger W., Dennis W. Bakke, and Roger F. Naill. *Creating Abundance: America's Least-Cost Energy Strategy*. New York: McGraw-Hill, 1984.

Scientific American. "Energy for Planet Earth." Entire September, 1990 issue devoted to energy.

World Commission on Environment and Development. "Energy: Choices for Environment and Development." *Our Common Future*. Oxford: Oxford University Press, 1987.

The following chapter in *Taking Sides: Clashing Views on Controversial Environmental Issues* discusses nuclear energy.

- Issue 8. "Is Nuclear Power Safe and Desirable?" (pp. 126–143)

Also see chapters on energy topics in the following editions of Worldwatch Institute's annual *State of the World* report.

- Christopher Flavin and Seth Dunn, "Reinventing the Energy System," *State of the World 1999*.

- Christopher Flavin and Nicholas Lenssen, "Designing a Sustainable Energy System," *State of the World 1999*.

- Christopher Flavin, "Harnessing the Sun and the Wind," *State of the World 1995*.

- Christopher Flavin and Nicholas Lenssen, "Reshaping the Power Industry," *State of the World 1994*.

- Marcia D. Lowe, "Rethinking Urban Transport," *State of the World 1991*.

- Marcia D. Lowe, "Cycling into the Future," *State of the World 1990*.

- Michael Renner, "Rethinking Transportation," *State of the World 1989*.

- Christopher Flavin, "Creating a Sustainable Energy Future," *State of the World 1988*.

- Christopher Flavin, et al., "Raising Energy Efficiency," *State of the World 1988*.

- Cynthia Pollock Shea, "Shifting to Renewable Energy," *State of the World 1988*.

- Christopher Flavin, "Electrifying the Third World," *State of the World 1987*.

- Christopher Flavin, "Reassessing Nuclear Power," *State of the World 1987*.

- Christopher Flavin, "Moving Beyond Oil," *State of the World 1986*.

Organizations

The United States uses more energy per person than any other society on earth. Like most Americans, you have many opportunities to make your home more energy efficient, and your life less energy-wasteful. Your local utility may have public programs to help customers make conservation improvements. Or you can contact one of the following organizations for more information about energy-efficient appliances, household energy use, and the broader role of energy conservation in national and international government policies.

The American Council for an Energy-Efficient Economy
1001 Connecticut Avenue NW, #801
Washington, DC 20036
Tel: (202) 429-8873; web: www.aceee.org

American Solar Energy Society
2400 Central Avenue, G-1
Boulder, CO 80301
Tel: (303) 443-3130; web: ases.org/index.html

Rocky Mountain Institute
1739 Snowmass Creek Road
Old Snowmass, CO 81654-9199
Tel: (970) 927-3851; web: www.rmi.org

UNIT 9

Save the Earth—
Feed the World

JILL SINGER

Planting trees in alternate rows with food crops, shown here in Kenya, helps prevent soil erosion because the trees' roots anchor the soil and recycle plant nutrients. This traditional farming practice, known as agroforestry, is currently being revived by new research.

BEFORE YOU VIEW THE TELEVISION PROGRAM

Learning Objectives

After completing the assigned readings and viewing "Save the Earth—Feed the World," you should be able to

- discuss the main environmental problems associated with heavy use of agrochemicals in farming in the United States.

- use examples from the program to explain the differences between industrial farming methods and traditional farming methods.

- discuss the "green revolution" approach to increasing food production and explain the problems associated with it.

- explain Indonesia's decision to ban 57 chemical pesticides and discuss the integrated pest management (IPM) approach now encouraged in that country.

- discuss the farming practices that cause accelerated soil erosion and explain why erosion poses a risk to food production.

- use examples from the program to show how low-income farmers can conserve water and nutrients and protect their fields from erosion.

- summarize the debate between advocates and opponents of a large-scale shift to low-input agriculture in the United States.

- explain how biotechnology may help increase food production and why some analysts expect its contribution to be limited.

Reading Assignment

Choose the material from either textbook as your reading assignment. Your instructor might assign additional readings as well.

Living in the Environment

Chapter 5, "Nutrient Cycles and Soils," Section 5-8

Chapter 12, "Food Resources"

Chapter 13, "Water"

Chapter 14, "Minerals and Soil Resources," Sections 14-6 and 14-7

Chapter 21, "Protecting Food Resources: Pesticides and Pest Control"

Environmental Science

Chapter 12, "Water Resources and Water Pollution," Sections 12-1 through 12-4

Chapter 13, "Mineral and Soil Resources," Sections 13-6 through 13-8

Chapter 15, "Food Resources"

Chapter 16, "Protecting Food Resources: Pesticides and Pest Control"

Unit Overview

This unit of *Race to Save the Planet* examines the environmental impacts of farming in industrial and developing countries and explores the prospects for increasing food production to meet the needs of a growing population without causing unacceptable environmental damage. Agriculture employs more people and occupies more land worldwide than any other human activity, and its impact on the global environment is correspondingly large. Environmental damage caused by farming is especially important because in addition to pollution, health risks, and disruption of natural ecosystems, it threatens the resources on which future harvests depend.

The first two chapters in the reading assignment introduce the fundamental resources on which food production depends and discuss how soil and water resources are managed by farmers to maintain and expand harvests. The text examines the causes and consequences of mismanagement of soil and water and discusses the sustainable management of these critical resources. The chapter on food resources introduces the major crops and farming methods used to sustain humanity, discusses problems of food production and distribution, and reviews methods of increasing harvests and making farming practices more sustainable. The chapter on pesticides and pest control summarizes major types of pesticides, reviews controversies surrounding their use, and presents alternative methods of pest management and control.

The accompanying television program, "Save the Earth—Feed the World," examines how industrial farming as practiced in Western countries developed from the stable and sustainable, but less productive, practices of the past. It considers the environmental consequences of overuse of the inputs—irrigation water, fertilizer, pesticides—that make modern farming so productive and reviews the experience of one developing nation, Indonesia, that has cut back on pesticides and adopted an innovative strategy for pest control known as integrated pest management. The program examines mismanagement of land resources and the problem of erosion in the United States and Australia, then visits West Africa to show how subsistence-level farmers can benefit by conserving soil, water, and plant nutrients. A profile of U.S. farmers that use alternative, low-input approaches sets the stage for the debate between advocates and skeptics of low-input sustainable agriculture. The program concludes with a look toward farming's future, reviewing the prospects for low-input approaches and the role that genetic engineering and other biotechnologies may play in producing food without undermining the land's productive potential.

The program title highlights the two-fold challenge facing farmers: how to preserve the land's productive potential while producing the food needed by a rapidly growing population. Farmers are discovering that by using inputs of water, chemicals, and energy as efficiently as possible they can save money and control environmental impacts while maintaining high productivity. Low-input agriculture combines the stability and sustainability of traditional practices such as mixed farming and crop rotations with the latest technologies and insights from current scientific research. This blend of old and new may provide the resilience needed to ensure food supplies in an age of global changes.

Glossary of Key Terms and Concepts

The following terms and concepts will be helpful as background for viewing "Save the Earth—Feed the World."

Farmers use many **agrochemicals** to increase the productivity of their crops, including synthetic fertilizers to increase crop yields and pesticides designed to kill insects and weeds. These chemicals, which make high yields possible, are also responsible for some of the most serious environmental impacts of modern agriculture.

Agroforestry is the general name for farming practices in which trees and crops are grown in combination. Agroforestry systems benefit from the ability of trees to protect soil from erosion and to capture and recycle plant nutrients.

Alley cropping is one method of agroforestry, developed in West Africa, in which food crops are grown between rows of fast-growing trees whose leaves and roots enrich the soil with nutrients.

Biotechnology encompasses a wide range of techniques used to manipulate living organisms to develop or accentuate characteristics that human beings desire. Genetic engineering, in which the hereditary material of a plant or an animal is modified at the molecular level, is one contemporary form of biotechnology with many applications in agriculture.

The **brown planthopper** is an insect that commonly damages rice crops in Southeast Asia. It became a serious threat to rice harvests in Indonesia when it developed resistance to the chemical pesticides most widely used there.

The **carrying capacity** of a pasture or range used for grazing is the number of cattle or other livestock it can support without suffering degradation or diminishing its productivity.

Simple stone **checkdams**, constructed along contours in a field, intercept rainfall runoff and slow its downhill flow to allow more water to soak into the root zone of crops. The method helps farmers in West Africa make the most of limited rainfall.

The traditional practice of **crop rotation,** or planting different crops in the same field in subsequent years in a prescribed pattern, helps maintain soil fertility and prevents populations of pests of any particular crop from growing to a size capable of doing extensive damage.

The **green revolution** refers to the package of high-yielding seeds, fertilizers, and pesticides introduced into some countries in Asia and Latin America in the 1960s and 1970s that resulted in dramatic increases in food production where favorable conditions (water for irrigation, credit for loans, farmers familiar with the cash economy) existed.

Families on the Indonesian island of Java (and other parts of Southeast Asia) raise intensive "**home gardens**" in which vegetables, fruits, and other plants that supply food, medicine, and material needs are combined in a system with the appearance and many of the properties of a natural ecosystem.

Integrated pest management (IPM) uses a combination of biological, chemical, and mechanical methods to control populations of crop pests so they do not increase to a size that can cause economically significant crop damage.

Low-input agriculture refers to farming methods that minimize reliance on fertilizer, pesticides, and other purchased materials by using these "inputs" as efficiently as possible. Organic farming (see below) is one type of low-input agriculture.

Minimum tillage refers to cultivation practices designed to minimize soil disturbance from plowing in order to control erosion and save energy.

Industrial agriculture typically depends on **monoculture**, in which a single uniform crop is grown on large areas. Monocultures are especially susceptible to pest outbreaks and disease and require costly methods of pest control.

Organic farming relies on natural sources of plant nutrients (manure, crop residues, legumes that build nitrogen into the soil) and natural methods of insect and weed control, rather than applications of agrochemicals. Organic farming is one form of low-input agriculture.

Salinization refers to an accumulation of salts in topsoil caused by evaporation of excessive irrigation water, a process that can eventually render soil incapable of supporting crops.

Shifting cultivation is a method of subsistence farming in which a field is cleared of woody plants, often by burning, then planted to crops for four to five seasons until the soil nutrients are depleted. The field is then abandoned for as long as 30 years to allow ecological succession to restore nutrients to the topsoil. If the cycle is repeated too quickly, nutrients are not restored, soil erodes, and crops fail on the depleted land.

Water harvesting is a traditional method used by farmers in arid West Africa and the Middle East in which the land surface is shaped to capture sparse rainfall and concentrate it on small growing fields to provide maximum moisture to crops.

Farmers in some parts of the world plant trees as **windbreaks** on the boundaries of their fields, to slow wind speeds and prevent soil erosion. Windbreaks were planted on the U.S. Great Plains after the disastrous Dust Bowl of the 1930s, then mostly abandoned in later decades as farmers again sought to maximize production from their land.

AFTER YOU VIEW THE TELEVISION PROGRAM

Consider What You Have Seen

Farmers in every country and at every economic level face the same concerns: how to combine soil, water, and seeds effectively to produce sufficient food and income and to maintain the integrity of the land. Farmers have just two kinds of resources available for this task: **internal** resources— sunlight, soil, locally available water, the genetic potential of their crops, and their own labor—and **external** resources—fertilizers, pesticides, hybrid seeds, fuel for farm equipment, and irrigation water from distant sources, all of which must be purchased off the farm. Subsistence farmers depend almost entirely on internal resources to produce their harvests, whereas the productivity of industrial agriculture rests on heavy use of external resources. The program shows examples of many different ways to adjust the balance between internal and external resources. Expanding production and protecting the environment are not necessarily mutually exclusive, but they are not always easy to achieve at the same time.

After watching the program, relate the material back to the following general themes of food production.

- Soil: The foundation of farming

- Water

- Farming methods and crops

- Controlling insects and weeds

- Choosing sustainable agriculture

Soil: The Foundation of Farming

Soil is more than a growing medium: Recall from your reading assignment that soil is a complex and varied resource, compounded of living and nonliving components. Review the discussion in your text about major soil types, and see if you can name the most likely type(s) of soil in some of the locations visited in the program. The predominant soils in West Africa, for example, have little humus and are sandy. Why would these characteristics make water management a crucial concern for West African farmers?

The best farming areas worldwide are underlain by deep nutrient-rich soils that sustain healthy and abundant crops. The Darling Downs in Australia is one such area, and the corn belt of the U.S. Midwest is another. These areas have such high natural fertility that even high losses of soil through erosion have little effect on crop production and can be easily offset by applications of fertilizer. Farmers in such areas may be tempted by a false sense of security to believe that their harvests are not at risk and that careful conservation measures are a needless expense.

Fertile soil is a renewable resource, regenerated by a combination of geological and biological processes. But it is not an infinite resource, and soil formation does not take place on a time scale meaningful to humans; as your text points out, it can take as long as 1,000 years for natural processes to rebuild an inch of topsoil. Every type of farming shown in "Save the Earth—Feed the World" depletes soil to some degree. But some farming methods emphasize retention and regeneration of the soil; they place a high value on this primary internal resource.

The faces of soil erosion: The television program contains examples of all the types of erosion discussed in your text. You should recognize that erosion is a natural process, a gravity-based tendency of loose particles to move downhill with wind and water. Accelerated erosion respects no economic boundaries; it is as common on profitable, wealthy farms in North America as on subsistence plots in developing countries. Identify the types of erosion shown in the program:

- Wind erosion—shown in Australia's Darling Downs, historical footage from the U.S. Great Plains during the Dust Bowl, and in Niger in West Africa.

- Sheet erosion—a special risk in West Africa, where the seasonal rains come in unpredictable, torrential downpours that can scour the surface of a slightly sloping field.

- Gully erosion—the most visually dramatic form of erosion by water, here shown in Australia's Darling Downs.

- Desertification from overgrazing—a combination of soil erosion and biological deterioration of grazing land, caused by too many animals. Composition of range vegetation shifts toward unpalatable and less nutritious grasses; soil is damaged by compaction and is exposed to wind and water by the constant pressure of animals. The potential productivity of the land—its carrying capacity—is gradually diminished. Clearly evident on grazing lands in Queensland, Australia, and in West Africa.

Conserving topsoil: Just as erosion is caused by farmers at all economic levels, methods of soil conservation depend on the same principles wherever they are practiced. One part of the strategy is to slow the flow of air or water. The windbreaks of the Majjia Valley in Niger brake windspeeds and reduce the potential for dustbowl-like conditions. The checkdams and water harvesting structures shown in Burkina Faso's Yatenga Plateau retard runoff; when water moves more slowly, it drops some of the soil it would otherwise carry in solution. These types of soil conservation practices attempt to control the medium that does the eroding.

The best protection for topsoil is to keep the ground covered with vegetation; roots hold the soil in place, and leaves protect it from wind and water. Only a few types of farming keep vegetation on the land at all times; Fred Kirschenmann's farm in North Dakota shows one approach to protecting fallow land with cover crops.

Land that is steep or sloping is especially prone to erosion. Farmers in steep or mountainous areas have developed many ways of keeping erosion in check. The construction of terraces, such as Indonesia's terraced rice paddies, have kept parts of Southeast Asia productive for centuries. But terraces are difficult to construct and maintain, and they are constructed by taking some land out of production. On more gently sloping land, like that shown in North Dakota, plowing along the land's natural contour is used to accomplish the same thing.

Soil conservation in any form can be an expensive undertaking, sometimes beyond the means of the individual farmer. It may involve changing a farmer's methods or taking some land out of production to plant a windbreak or construct a water-catching terrace, and it may require more labor. Soil conservation can sometimes seem to be at odds with a farmer's immediate economic interests, but it is generally in society's best interests. The Naam movement in Burkina Faso, started by Bernard Ouedraogo, shows one way that society's interests can be marshaled to tackle conservation work beyond the means of any individual farmer.

Water

Water is the lifeblood of agriculture; naturally available water, supplied through the hydrological cycle, is a key internal resource for farmers everywhere. But as is obvious from your readings and from the examples shown in "Save the Earth—Feed the World," fresh water is not distributed equitably. In some areas it is far too scarce to support farming; in others its overabundance threatens agriculture. Farming systems around the world show a range of solutions to water problems; supplying water for agriculture also causes some of the most serious environmental problems associated with farming.

The hydrological cycle: The climate patterns that determine rainfall shape the world's agricultural practices. Review Unit 2 if you have forgotten how the sun's energy controls the water cycle. The television program visits locations characterized by the extremes of the water cycle: drought-prone West Africa, where rain falls only during a short rainy season, and humid Indonesia, where rain falls throughout the year and water is rarely scarce. Parts of the temperate zone get moisture throughout the year, but the pattern of seasons means that some precipitation (snowfall) is not immediately available to support plant growth. The historical farming practices in Sturbridge Village, Massachusetts, based on methods developed in Europe, show a farming system adapted to this seasonal pattern of water availability and a summertime growing season.

Irrigation: The artificial delivery of water to crops is enormously important to food production. Water is the key limiting factor to agricultural production in many parts of the world, and transferring surface water or pumping groundwater increases harvests of many crops. Only about 10% of the world's cropland is irrigated, but this area accounts for a third of world food production.

Like soil, irrigation water is often mismanaged and commonly used wastefully. Yet such waste can have serious environmental consequences. One discussed in the television program is soil contamination by salinization in Australia's Murray Darling basin. Review the description of salinization in your textbook. Salt buildup, which can reduce yields and even cause land abandonment, is a problem virtually everywhere that irrigation is practiced. List some of the responses to salinization.

The water-harvesting methods practiced by farmers in Burkina Faso show how farmers under difficult conditions can make the most of a limited resource. Rainfall in this part of Africa is erratic and often doesn't soak the soil thoroughly enough to support a good crop. Water harvesting lets farmers concentrate water collected from a large area onto a small growing area, ensuring that the soil is moistened far below the surface. In this system, water is treated like a precious, irreplaceable resource—it can mean the difference between a good crop and a failed harvest.

Why aren't all farmers so frugal with their water? Often it is a matter of economics. Most large irrigation projects are constructed by governments, and farmers who use the water pay a low subsidized price—often lower than the cost of distributing water to them through the system of dams and canals. Because water's true economic value is hidden in such cases, farmers have no economic incentive to conserve it. It is perhaps ironic that Burkina Faso's subsistence farmers, whose rain-fed water supply is in effect "free," treat water as the most valuable substance on earth.

Water as a medium for contaminants: Water, of course, is not solely a resource for food production. It is an agent of erosion, as we have already discussed, and it is a medium that can be contaminated. Agriculture is the greatest polluter of water on earth—a nonpoint source of nutrients such as nitrogen and phosphorus and of synthetic chemicals. The liberal use of agrochemicals in industrial agriculture means that many of these chemicals—fertilizers, insecticides, herbicides—don't reach the plants they are intended to nourish or protect. They are washed into drainage ditches and end up in surface water, or they are washed through the soil itself and end up in groundwater supplies. The contamination of water may be the most serious environmental problem associated with farming.

Recall the examples of groundwater contamination discussed in the program. Are the chemicals responsible for this contamination an internal or an external resource of California farms? Why do you think farmers might fail to control agrochemical contamination of water supplies? What could be done to correct the situation?

Farming Methods and Crops

Your textbook points out that there are no more than a few dozen major food plants cultivated by humanity, and of these a tiny handful provide most of the world's food. That is not because the rest of the 248,400 named plant species are inedible but because their nutritional properties and suitability for cultivation have not yet been explored. The major crops shown in "Save the Earth—Feed the World" include rice, the basis of diets in Indonesia and throughout Asia; wheat, a major grain crop consumed directly and in baked goods; corn, which is a food staple in many developing countries and an important source of feed for livestock in industrial countries; and sorghum and millet, grain crops grown primarily by subsistence farmers in semiarid areas including the West African Sahel region. Most of these crops were domesticated from wild grasses and cultivated when societies first settled in agricultural communities, 10,000 to 12,000 years ago.

Other foods shown in the television program include the fruits and vegetables grown in Indonesian home gardens, and the vegetables grown in California's Central Valley. Important as these crops are as sources of vitamins and dietary variety, they make only a small contribution to the total calories consumed by human beings.

Not shown in the program, but of great importance to world agriculture, are the many crops not consumed directly as a source of calories, such crops as cotton, coffee, and tea. These crops are often major exports from developing countries, and they are often grown on land that could otherwise be planted to food crops. In a sense, they are indirectly a part of the world's food supply. They generate income that can be used to purchase food, and they occupy land that could, if needed, be used to produce food.

Types of farming: The program shows a good cross section of the farming practices characteristic of agriculture today. Large-scale industrial agriculture, heavily dependent on mechanized equipment and purchased ("external") inputs of fuel and chemicals, is illustrated by examples from the United States and Australia. Rice production in Indonesia shows how this industrial approach has been introduced to a developing country, where the scale of production is smaller and more intensive but the methods of production are still largely based on external resources. The program also gives several contemporary examples of subsistence agriculture, including the intensive Indonesian home gardens, the examples of shifting cultivation from West Africa and Southeast Asia, and the more settled farming practices in West Africa. Most subsistence-farming systems are based on the productive potential of internal resources.

How do farmers increase production?   Perhaps the greatest challenge facing agriculture is how to continue to expand food production to keep pace with a growing human population. There are only a small number of ways to increase production, and each has different implications for the environment. The most obvious way was, until the middle of this century, the most important: Plant more land. Today, about 12% of the earth's land surface is devoted to agriculture, and no frontier remains from which new cropland could be cleared. Remaining areas are less fertile, steeper, drier, and in other ways less suitable for crop production. Some new land is brought into cultivation each year, but today this addition is largely offset by land abandoned due to erosion, salinization, low productivity, and other forms of environmental deterioration.

The other major method of expanding food production is increasing the harvest from existing farmland—raising crop yield, the quantity of a crop harvested per acre or hectare (1 hectare is equal to approximately 2.5 acres). This is the approach that has been responsible for the unprecedented growth in world food production since the end of World War II, and it has been accomplished by increasing the use of external inputs in agriculture.

Indonesia's experience in boosting rice pro-

duction, described in the television program, illustrates this approach. Raising yields by increasing use of fertilizer, pesticides, and hybrid seeds is the essence of the so-called green revolution. The rice varieties that were the key to Indonesia's growing harvests were most productive in combination with fertilizer and chemical pesticides. The country successfully raised harvests, enough to become virtually self-sufficient in rice, but did so at a price: Indonesia's farmers became dependent on inputs that had to be purchased.

The Indonesian example seems to suggest that the only way to raise crop yields is to increase reliance on external inputs, to channel more agrochemicals into food production. When yields are at subsistence levels, this may be true. But as we noted previously, many of the chemical inputs applied to crops don't reach their target plants but are washed away in soil or runoff. If the inputs could be delivered more efficiently to plants, and waste reduced, it might be possible to increase yields without increasing use of agrochemicals and irrigation water. This concept of efficiency, parallel to the rationale for energy efficiency, lies at the heart of low-input, sustainable farming methods.

While it may be possible to increase crop yields by using external resources more efficiently, scientists are also working on the possibility that crops themselves can be made more efficient—that they can perform better than existing crop varieties. "Save the Earth—Feed the World" discusses the possibility that biotechnology may offer new ways of increasing crop yields and getting more food production from each unit of land.

Energy—the "invisible" input in industrial agriculture: Though the television program does not discuss it in detail, you should recognize the important role that energy plays in modern industrial agriculture. Energy, particularly in the form of fossil fuels, has played a central role in the growth of world food production since mid-century, and possible limits on fossil fuel use to slow global climate change have important implications for farming.

Obviously, all agriculture depends on the capture of the sun's energy in living plants. But in addition to this "internal" energy, purchased sources of energy are important to industrial farming. The direct uses include the fuel needed by farm equipment—plows, cultivators, and so on—and energy used to pump irrigation water (generally either diesel fuel or electricity). Indirect uses are also important: Chemical pesticides are generally made from oil-based chemical feedstocks, and nitrogen fertilizer is manufactured from natural gas. The cost of these energy inputs is determined by fuel costs, and food prices are thus influenced by the cost of energy. Try to identify the direct and indirect energy uses in farming in Indonesia and in Australia, in locations shown in the television program.

A glimpse of sustainable agriculture: The television program considers several examples of farming methods more sustainable and less damaging to the environment than are widely prevailing practices. Two examples are Fred Kirschenmann's organic farm in North Dakota and the use of integrated pest management approaches to control the brown planthopper in Indonesia. Contrast each of these examples with your text's discussion of the characteristics of modern industrial agriculture and conventional means of pest control using chemical pesticides. What are the principal sorts of environmental impacts that these two examples reduce or avoid?

Controlling Insects and Weeds

If food production were merely a matter of bringing together sunshine, water, nutrients, and seeds, food production for an expanding population might not be considered a serious challenge. But humanity faces stiff competition for food supplies. Weeds compete with crop plants for sunlight and nutrients, reducing crop yields. Insects damage crop plants and can decimate harvests. Since the first agrarian communities were established in the Middle East, humanity has battled pests for the control of harvests, and in this century the battle has generated some of the most bitter controversies and severe environmental problems associated with farming.

The industrial approach to pest control, based on chemical pesticides, is a product of our times and was instrumental in stimulating the environmental awakening in the United States in the early 1960s. Recall from Unit 3 the role played by Rachel Carson's book *Silent Spring*. The controversy over chemical pesticide use alerted many people to the fact that agriculture was not wholly benign. Since Carson's book was published, however, the number and variety of pesticides in use have increased dramatically, and the unwanted consequences have multiplied. Pest control is undeniably necessary in agriculture. The question is how best to achieve it.

Why do pest problems persist? "Save the Earth—Feed the World" illustrates the features of contemporary food production that virtually guarantee an ongoing battle with pests. In temperate-zone countries such as the United States and Australia, industrial farming relies on large, uniform plantings of crop monocultures—an unlimited food supply for crop-dependent insects. In tropical countries, the combination of green-revolution–style monoculture plantings with year-round warmth makes conditions even more favorable to pests. The dense plantings necessary to support human populations are inevitably at risk.

A further problem is that both competitive weeds and crop-damaging insects reproduce quickly and their populations can develop resistance to chemical agents used to control them. Review the discussion of resistance in your textbook, and make sure you understand how and why chemical agents are almost inevitably limited in their ability to control biological phenomena.

Pest control is a necessary evil: As long as the human species relies on agriculture, pest control will remain a problem. Farming involves artificial, simplified, heavily managed ecosystems. Such systems are inherently unstable and lack many of the features of natural ecosystems that keep pest populations in check. To illustrate this point, think of the difference between the Indonesian home gardens and the Indonesian rice paddies shown in the program. Which system is less likely to suffer serious damage from pests? Why? What elements of a natural ecosystem contribute to keeping pest populations in check?

Though pest control is necessary, there are many ways to achieve the goal of reducing pest populations below levels that can cause serious damage. Using chemicals designed to kill insects or weeds, either indiscriminately or selectively, is only one such approach. Building some of the features of natural ecosystems into farming systems is another.

The consequences of overuse: Excessive use of pesticides has both ecological and human consequences. The television program emphasizes the human consequences, which are serious in their own right and also influence public attitudes about industrial agriculture.

The human consequences include the effects of both direct and indirect exposure. Consider the case of the farmworkers in California's Central Valley whose children show an unexpectedly high incidence of cancer. Pesticides are designed to disrupt biological systems, and some pest-killing chemicals are known to cause animal cancers. But a direct link to human cancers in a case like this, though intuitively obvious, is difficult to prove. From a medical standpoint, the case raises some of the same problems as does determining the health effects of exposure to air or water pollution, discussed in Unit 5. Direct exposure to pesticides is undeniably dangerous. But how dangerous? What should the standards for allowable exposure be?

As mentioned earlier, pesticides are among the external inputs most commonly washed into surface water and groundwater supplies, where their presence raises additional concerns because it multiplies the population indirectly exposed to the chemicals. Pesticides have been detected in groundwater in every farming area in the United States and in many other parts of the world. Groundwater provides at least some of the drinking water in virtually every community in the country. What risk is posed by the presence of small concentrations of pesticides in water supplies?

Another place that pesticides end up, in small concentrations, is in residues on food supplies. In recent years, questions about the safety of food supplies have raised public awareness and concern about pesticides as never before. All of these levels of concern are motivated by fears about agrochemicals that end up where they are not wanted: in the bloodstream of farmworkers, in drinking water supplies, on the skin of fruits and vegetables at the supermarket. Are unmeasurable risks and elusive cause-and-effect relationships the inevitable price of the modern food system?

The changing science of pest control: "Save the Earth—Feed the World" points to another way. The approach known as integrated pest management (IPM), described in your textbook, applies ecology to what is basically an ecological problem.

Recall the discussion of Indonesia's IPM program, and try to list the various methods combined in this program. IPM uses a combination of means, some of them biological, chemical, and mechanical, to control pests. The philosophy is very different than the attitude toward pest control that has prevailed for the past generation. IPM is based on the idea that there is no need to eliminate pests entirely, if their populations can simply be kept below a threshold size at which they cannot cause economically significant damage to a crop. Achieving this goal is not easy. But experience in Indonesia, the United States, and many other places around the world proves that it is possible. Combined with other changes in farming practices, IPM is a way to reduce many real and suspected risks of high-input farming.

Choosing Sustainable Agriculture

"Save the Earth—Feed the World" highlights some of the choices that must be made to make farming more sustainable and address its most serious environmental problems.   The choices facing farmers and societies involve where to strike the balance between the internal and external resources on which food production depends. It has become clear that external resources—fertilizers, pesticides, and so on—cannot compensate fully or permanently for the degradation and loss of internal resources, such as fertile soil.

Attributes of low-input, sustainable agriculture include frugality and efficiency in the use of purchased inputs, careful management of soil and water, and the blend of pest control methods characteristic of IPM. In the new thinking about farming, internal and external resources are not seen as substitutes but as complementary resources. The philosophy applies as much to limited-resource farmers in the West African Sahel, seeking to make their modest harvests less vulnerable to catastrophe, as to large-scale farmers in the U.S. Midwest who look for ways to cut back on chemical and fuel use as cost-saving measures. The approach blends contemporary scientific knowledge with the insights of traditional farming systems and the patterns of natural ecosystems.

This approach to farming does much to address the most serious environmental problems associated with present-day agriculture. But that is only half the challenge for the future. Expanding food production rapidly enough to meet the needs of a doubling human population will require higher yields—more production from each acre of cropland. That will not be as easy to accomplish as it was in the past, partly because the areas most suited to intensive, industrial-style farming have already been exploited and partly because concerns about the environmental consequences of high-input farming have led to a reappraisal of the main ways used to raise yields in the past. Some scientists hope that biotechnology will help plant breeders develop new crop varieties that can produce higher yields without heavy use of agrochemicals.

But as the program points out, a great deal of uncertainty surrounds biotechnology. Its benefits to farming are not expected to come any time soon. The methods of modifying plants with biotechnology are still primitive, and applications to widely grown crops are not yet available. In addition, there is no guarantee that biotechnology will reinforce the shift toward low-input approaches in farming. Some researchers are now working on ways to give plants characteristics, such as resistance to pest damage, that chemicals now provide. This work, like the research on rice at the Salk Institute shown in "Save the Earth—Feed the World," could be compatible with low-input, sustainable agriculture. But agrochemical companies, which also support biotechnology research, hope to develop crop varieties that are more resistant not to pests but to the herbicides they sell, so that more of these chemicals could be applied to farmers' fields. This approach has been widely criticized as incompatible with solving the main environmental problems facing agriculture.

The use of a new biotechnology, like all technologies in agriculture, will depend on the priorities of those who develop the technology. Its impact on food production will depend on how the technology is made available to farmers and whether it meets their needs. The answers to such questions depend on public policies, on governments, and on the leadership and pressure provided by citizens.

Many questions about the future of farming are still unresolved. But the new enthusiasm for low-input approaches signals an important shift in attitudes about farming, both among farmers and the general public. The belief that there will always be more land to plow is disappearing, as resource

limits become clear. Societies are beginning to move more toward *stewardship* of their food-producing resources—careful management of the internal resources of agriculture and efficient and responsible use of the external resources used to increase output. Both economic and environmental concerns reinforce this shift. Do you think that "Save the Earth—Feed the World" made a good case for the need for low-input approaches? What questions does the program leave unanswered?

Take a Closer Look at the Featured Countries

Australia, which we first visited in Unit 4, is one of the world's major food-producing countries and an important source of grain exports on the world market (Figure 9-1). An affluent nation with just 19 million people, Australia depends on industrial agriculture with farms and ranches often covering enormous areas. The first European settlers arrived only two centuries ago, and Australians still tend to value a "frontier spirit" that places little emphasis on careful management of soil and water resources. The resource problems shown in "Save the Earth—Feed the World," while by no means universal, are commonplace in Australia's major crop and livestock areas.

With a subtropical and temperate, but largely arid, climate, Australia's major crop is wheat. In 1988, the country's total grain production was about 21.2 million metric tons, or more than 1.3 tons for each citizen. This is much more than is needed for direct food consumption and means that much of the harvest can be exported. In recent years more than 60% of Australia's harvest has been sold on the world market. The country plants 13.8 million hectares (about 35 million acres) to grain crops and harvests an average of 1.5 tons per hectare. This yield level, not particularly high, is a reflection of Australia's extensive farming methods and its reliance on wheat. Wheat is not usually heavily fertilized and produces less grain per hectare than does corn or rice.

The vast scale of Australian farms and ranches and the heavy reliance on grain exports have both tended to work against careful soil and water management. Farmers anxious to expand harvests

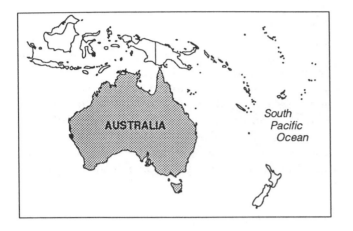

Figure 9-1   Australia

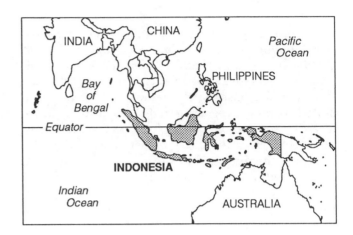

Figure 9-2   Indonesia

sacrifice conservation measures. But dry years in the late 1980s has reduced the country's harvests and exports and helped focus public attention on the condition of Australia's land resources. As shown in the television program, there are signs that a new attitude toward land management is emerging in Australia, prompted by evidence of cumulative damage caused by past practices.

Indonesia, an island archipelago north and west of Australia in Southeast Asia, is one of the world's most populous countries with 212 million people (Figure 9-2). The population is growing at about 1.6% per year, somewhat slower than the average for countries in Southeast Asia but still fast enough to double the population in 44 years. A longstanding government commitment to family planning helped bring the growth rate down to its present level and is likely to reduce it further.

Parts of Indonesia, for example the island of Java, are among the most densely settled regions on earth, and providing sufficient food for the large and growing population is the country's greatest challenge. The staple crop is rice; in 1988, Indonesia harvested 28 million tons of rice from 9.8 million hectares (24 million acres) of paddies and fields, or about 156 kilograms (343 pounds) for each Indonesian. This is just barely enough to cover the country's subsistence needs, and the country is considered self-sufficient after long being dependent on food imports. Indonesia still imports some rice to maintain adequate supplies; in recent years, imports have varied from as little as 28,000 tons in good years to more than 400,000 tons in years with poor harvests. Each year the country

must produce or purchase enough additional food to support 3 million new Indonesians.

The country's rice production has more than doubled since 1970, while the area planted to rice increased by only 20%. The gains have come from adopting green-revolution approaches and applying fertilizers and pesticides to new high-yielding rice varieties. The abundance of water allows farmers to make the most of these new inputs. Rice yields are now more than 2.8 tons per hectare, a level much higher than in most developing countries. The productivity of farming reflects both the widespread use of agrochemicals and the country's natural endowment of water and highly fertile volcanic soils. But the green-revolution approach has also brought problems, such as increased vulnerability to outbreaks of pests like the brown planthopper that have developed resistance to chemical pesticides. Indonesia needs the productivity of modern farming methods but cannot afford vulnerability to large-scale crop failures.

Many traditional farming practices in Indonesia, including the home gardens of Java shown in "Save the Earth—Feed the World" and some shifting cultivation methods once common in the country's forest areas, were stable and sufficiently productive when population pressure was low. The challenge to food production in Indonesia is how to incorporate features from these traditional practices into modern cultivation systems so that they can produce enough food for a large and growing population without serious risk of environmental contamination or widespread crop damage.

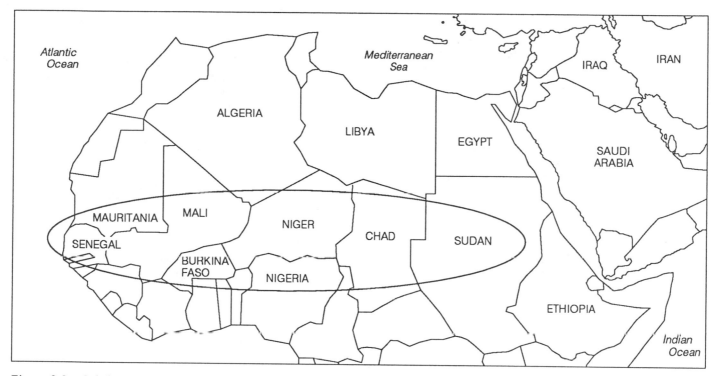

Figure 9-3   Sahel region of Sub-Saharan Africa

Few regions on earth are more hostile to food production than is the **Sahel region of Sub-Saharan Africa**, which includes the countries of Burkina Faso and Niger shown in the television program. "Sahel," an Arabic word meaning coast or border, names a band of arid land south of the Sahara desert from Senegal and Mauritania in the west to Sudan in the east (Figure 9-3). The entire region is vulnerable to periodic droughts that may persist for years; the short rainy season brings moisture in torrential but spatially erratic downpours that may flood some areas while others remain parched. Soils are infertile, there are few available sources of irrigation water, and the region's farmers are far too poor to afford fertilizers or other costly inputs.

People in the Sahel region traditionally raised livestock, which could be herded to graze new growth supported by spotty rains. But settled farming has become more common in the Sahel as populations, and food needs, have grown. Expanding farmland and livestock herds have seriously increased pressure on the fragile Sahel environment, and drought often means catastrophic crop failures. Most farmers in the Sahel grow sorghum and millet, subsistence grains that can

survive the severe arid climate. Without irrigation or fertilizer and with unpredictable rainfall, such crops are not very productive; grain yields range from as little as 300 to no more than 800 kilograms per hectare.

Farmers under these conditions are often more concerned about reducing their vulnerability to catastrophe than about increasing their overall yields. The best ways to reduce risks involve careful management of the internal resources available to farmers: harvesting water to make the most of limited rainfall, protecting soil from erosion with windbreaks like those in Niger's Majjia Valley, and planting trees such as *Acacia albida*, whose roots build nitrogen into the soil, in among their crops. Many such practices are rooted in cultural traditions that were slowly being lost from the region due to contact with outsiders and pressure to enter a cash-based economy. But grass-roots movements, such as the Naam movement in Burkina Faso, profiled in the television program, promote small-scale development based on appropriate regional traditions.

By revitalizing a tradition of community work during the dry season, Naam founder Bernard Ouedraogo turned one of the Sahel's greatest liabilities—the long dry season and the underemployment at a time when no crops could be grown—into a resource of time and skill used to make subsequent rains yield more abundant harvests. Since its beginnings, the Naam movement has fostered local leadership, emphasized the contributions made by women and children to collective welfare, and restored and enhanced the land's productive capacity in one of Africa's most difficult environments. Such approaches cannot protect the people of the Sahel from the devastation of drought, but they can begin to restore resilience to Sahelian ecosystems and dignity to its communities.

The **United States** possesses some of the most productive farmland of any country in the world and uses industrial farming methods to produce a staggeringly large harvest (Figure 9-4). In 1989, the U.S. harvested 284 million tons of grain, primarily corn and wheat, from 63.2 million hectares of cropland (about 156 million acres). This was 1,141 kilograms for each person in the country. Average grain yields in the United States are high, nearly 4.5 tons per hectare—a level reflecting the country's fertile soils, reliance on highly productive crop varieties, and heavy use of fertilizer.

The United States is the world's largest exporter of grain, and roughly a third of the country's annual production is sold on the world market. Dozens of countries depend on imports from the U.S. for part of their food supplies. In recent years, the U.S. has exported nearly 100 million tons of grain each year, almost five times Australia's total grain production. Much of the grain consumed *within* the country is consumed indirectly in the form of meat; a large share of the U.S. harvest supplies feed for cattle, pigs, and poultry.

Despite its extraordinary productivity, U.S. agriculture is vulnerable in many ways, some of them shown in "Save the Earth—Feed the World." Large-scale monocultures in the corn belt and other growing regions are vulnerable to pest damage, and soil erosion is a serious problem. Runoff carrying nutrients from farmland is a major source of water pollution in most parts of the country, and groundwater reveals traces of pesticides everywhere it has been tested. The economics of farming in North America have made it difficult for many farmers with smaller operations to stay in business, and large-scale "agribusiness" agriculture has

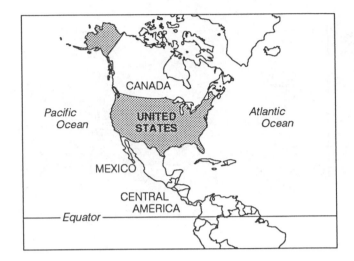

Figure 9-4   The United States

become the predominant model. In addition, computer models of the global climate suggest that some of the most productive farming regions in the country are likely to be adversely affected by climate change, with consequences that would have worldwide repercussions.

Interest in low-input farming methods is growing, though such methods are still used by only a small minority of the country's 2 million farmers. As those sustainable-farming advocates demonstrate that they can produce good harvests and make a profit with alternative methods, their ranks seem likely to grow. But as long as the United States places heavy emphasis on maintaining and expanding farm exports, many farmers will still favor practices that ensure maximum production now at the expense of soil, water, and other resources.

Examine Your Views and Values

1. Do you think that some contamination of food products or drinking water with pesticide residues is an acceptable price to pay for attractive and abundant produce? When you shop, are you more concerned with price or with quality? Would you buy fruit if it had been tested for pesticides and the level of residue, though measurable, was announced to be "safe"?

2. Do you think society, through the government, should pay farmers to adopt soil-conserving practices? Why or why not?

3. If you have friends or relatives involved in farming, ask their views about low-input agriculture. Why do they support it or oppose it?

TEST YOUR COMPREHENSION

Self-Test Questions
(*Answers at end of Study Guide*)

Multiple Choice

1. The farm at Sturbridge Village, which shows farming methods used in New England in about 1830, is the same as modern farming in the United States in all of the following ways *except* which one of the following?
   a. Crop rotation is used to maintain the fertility of the soil.
   b. Plants and animals are combined in a mixed farming system.
   c. The main source of plant nutrients is purchased off the farm.
   d. Crop residues and manure are returned to the soil.
   e. Fossil fuels are not used directly or indirectly in the system.

2. One of the main reasons given for expansion of U.S. food production in recent years has been
   a. the need to give free food to victims of starvation in developing countries.
   b. the opportunity to sell U.S. grain to buyers overseas.
   c. a desire to end hunger within the United States once and for all.
   d. careful land management that has made U.S. farmland more fertile.
   e. the increasing number of farmers in the United States.

3. One problem with aerial pesticide sprays mentioned in "Save the Earth—Feed the World" is
   a. the chemicals are quickly broken down by sunlight and become ineffective.
   b. they are widely opposed as a form of air pollution.
   c. the energy cost of aerial spraying cannot be justified by the value of the crops.
   d. only a small amount of the chemicals sprayed actually reaches the target crop.
   e. More than one of the above is true. The correct answers are _____.

4. The link between cancers among farmworkers in California's Central Valley and exposure to pesticides might be described as
   a. clearly conclusive.
   b. suspicious but inconclusive because such cancers are common.
   c. suspicious but inconclusive because the scientific data are incomplete.
   d. implausible because pesticides do not cause such cancers in humans.
   e. implausible because farmworkers build up an immunity to pesticides.

5. Traditional Indonesian home gardens are breaking down because
   a. a combination of population pressure and economic forces is leading farmers to specialize in cash crops and reduce the diversity of the gardens.
   b. the practice gradually but inexorably depletes the nutrients needed to support fruit and vegetable crops.
   c. the gardens have become more vulnerable to damage by the brown planthopper and harvests are suffering.
   d. introduction of the green revolution in Indonesia switched dietary preferences and increased demand for rice.
   e. they are more suited to large-scale production that is no longer possible in crowded Indonesia.

6. The main reason that Indonesia shifted to reliance on integrated pest management to control pests in the country's rice crop was that
    a. standard chemical pesticides don't work well with green-revolution–style farming methods.
    b. a new insect pest was introduced to the country that conventional pesticides could not control.
    c. a prominent environmentalist became minister of agriculture.
    d. a major pest of rice developed resistance to widely used pesticides and threatened the country's harvest.
    e. the United States, encouraged by success with IPM at home, introduced the method to Indonesia.

7. The response to the Dust Bowl in the United States during the 1930s included all of the following *except*
    a. planting windbreaks on field boundaries.
    b. strip cropping.
    c. fallowing practices.
    d. farmer education.
    e. agroforestry methods.

8. Soil erosion became a serious problem in the Darling Downs region of Australia because
    a. the region's soil was highly fertile and erosion seemed harmless.
    b. farmers in the region lacked the financial resources to prevent erosion.
    c. uncontrolled water runoff led to creation of severe gullies.
    d. opportunities to export wheat led farmers to emphasize production.
    e. More than one of the above is true.
        The correct answers are _____.

9. The Naam movement in Burkina Faso helps farmers in the Sahel region to
    a. adopt integrated pest management techniques.
    b. construct checkdams and other water-harvesting structures.
    c. learn better management of the internal resources available to them.
    d. reduce their reliance on purchased chemical fertilizer.
    e. More than one of the above is true.
        The correct answers are _____.

10. Fred Kirschenmann's farm in North Dakota is included in the program to show that
    a. organic farming, though it is stable, cannot compete with neighboring conventional farms.
    b. only farmers who have always farmed organically are likely to succeed with low-input approaches.
    c. farms using low-input approaches can be both productive and profitable.
    d. farms using low-input approaches can be productive but not profitable.
    e. even fertilizer-industry representatives are convinced by low-input farming.

11. The television program suggests that future increases in crop yields
    a. could be easily accomplished if farmers conserved soil and water.
    b. will depend on more farmers using available technologies.
    c. are likely to come from genetic engineering and other biotechnologies.
    d. could be a welcome fringe benefit of research on herbicide resistance.
    e. could be achieved if money is available for research on yields.

True or False

1. The green revolution is now practiced by most farmers in developing countries. _____

2. Problems with resistance to pesticides, though common in developing countries, rarely occur in countries like the United States. _____

3. Integrated pest management sometimes uses chemical methods. _____

4. Unlike Australia, the United States solved its erosion problems after the Dust Bowl experience. _____

5. The greatest problem facing farmers in the Sahel region is water management. _____

6. The risks of exposure to agrochemicals are well understood. _____

Sample Essay Questions

1. Contrast the farming practices demonstrated at Sturbridge Village, typical of New England 150 years ago, with prevailing farming methods today in the United States. What are the main differences?

2. What are the major environmental problems associated with heavy use of agrochemicals in industrial farming?

3. What was the green revolution? Use an example from the television program to explain the advantages and disadvantages of this type of farming.

4. Compare and contrast the types of farming practiced in Australia and in the Sahel region of Africa. Discuss the types of crops grown, the farming methods used, and the main environmental problems associated with each.

5. What are the two primary challenges facing agriculture? Do you think low-input agriculture provides an adequate response to both challenges? Why or why not? Use examples from the television program to support your answer.

GET INVOLVED

References

Bongarts, John. "Can the Growing Human Population Feed Itself?" *Scientific American*, March 1994.

Bray, Francesca. "Agriculture for Developing Nations." *Scientific American*, July 1994.

Brown, Lester R. *Full House: Reassessing the Earth's Carrying Capacity*. New York: W.W. Norton, 1994.

Brown, Lester R. *Tough Choices: Facing the Challenge of Food Security*. New York: Norton, 1996.

Brown, Lester R. *Who Will Feed China?* New York: Norton, 1994.

Care, James B. and Maureen K. Hinkle. *Integrated Pest Management*. Washington, D.C.: National Audubon Society, 1994.

Carson, Rachael. *Silent Spring*. Boston: Houghton-Mifflin, 1962.

Crosson, Pierre R., and Norman J. Rosenberg. "Strategies for Agriculture." *Scientific American*, Sept. 1989.

Eisenberg, Evan. "Back To Eden." *The Atlantic Monthly*, Nov. 1989.

Evans, C. T. *Feeding Ten Billion: Plants and Population Control*. New York: Cambridge University Press, 1998.

Gardener, Gary. *Shrinking Fields: Cropland Loss in a World of Eight Billion*. Washington, D.C.: Worldwatch Institute, 1996.

Grigg, D. B. *Agricultural Systems of the World*. Cambridge, England: Cambridge University Press, 1974.

Jackson, Wes, Wendell Berry, and Bruce Colman. *Meeting the Expectations of the Land: Essays in Sustainable Agriculture and Stewardship.* San Francisco, CA: North Point Press, 1984.

Lappé, Francis M. *World Hunger: Twelve Myths*. New York: Group, 1999.

National Research Council, Board on Agriculture. *Alternative Agriculture.* Washington, D.C.: National Academy Press, 1989.

Postel, Sandra. *Pillar of Sand: Can the Irrigation Miracle Last?* New York: Norton, 1999.

Reaganold, John P., et al. "Sustainable Agriculture." *Scientific American*, June 1990.

Wolf, Edward C. *Beyond the Green Revolution: New Approaches for Third World Agriculture.* Washington, D.C.: Worldwatch Institute, 1986.

World Commission on Environment and Development. "Food Security: Sustaining the Potential." *Our Common Future.* New York: Oxford University Press, 1987.

The following chapter in *Taking Sides: Clashing Views on Controversial Environmental Issues* takes a close look at the role of pesticides in food production.
   . Issue 9. "Is the Widespread Use of Pesticides Required to Feed the World's People?" (pp. 144–163)

Also see the following chapters in various recent editions of Worldwatch Institute's annual *State of the World* report that deal with farming and food production issues.
   . Lester R. Brown, "Feeding Nine Billion," *State of the World 1999.*
   . Lester R. Brown, "Struggling to Raise Crop Productivity," *State of the World 1998.*
   . Lester R. Brown, "Nature's Limits," *State of the World 1995.*
   . Gary Gardener, "Preserving Agricultural Resources," *State of the World 1996.*
   . Gary Gardener and Brian Halweil, "Nourishing the Underfed and Overfed," *State of the World 2000.*
   . Peter Weber, "Protecting Oceanic Fisheries and Jobs," *State of the World 1995.*
   . Anne Platt McGinn, "Promoting Sustainable Fisheries," *State of the World 1998.*
   . Megan Ryan and Christopher Flavin, "Facing China's Limits," *State of the World 1995.*

   . Lester R. Brown and John Young, "Feeding the World in the Nineties" *State of the World 1990.*
   . Sandra Postel, "Redesigning Irrigated Agriculture," *State of the World 2000.*
   . Sandra Postel, "Saving Water for Agriculture," *State of the World 1990.*
   . Sandra Postel, "Halting Land Degradation," *State of the World 1989.*
   . Edward C. Wolf, "Raising Agricultural Productivity," *State of the World 1987.*
   . Lester R. Brown, "Conserving Soils," *State of the World 1984.*

Organizations

Many nonprofit groups are active in promoting the shift toward low-input agriculture. The following list includes groups that conduct field research, carry out public education, and advocate public policies to reduce the environmental impacts of farming and encourage farmers in the United States to try low-input approaches. The national environ-mental organizations listed here can also suggest sources of information on sustainable farming in other countries around the world.

Institute for Alternative Agriculture
9200 Edmonston Road, Suite 117
Greenbelt, MD 20770-1551
Tel: (301) 441-8777; web: www.hawiaa.org

The Land Institute
2440 East Wellwater Road
Salina, KS 67401
Tel: (785) 823-5376; web: www.landinstitute.org

National Audubon Society
700 Broadway
New York, NY 10013-9501
Tel: (212) 979-3000; web: www.audubon.org/nas

Natural Resources Defense Council
40 West 20th St.
New York, NY 10011
Tel: (212) 727-2760; web: www.nrdc.org

Soil and Water Conservation Society
7515 Northwest Ankeny Road
Ankeny, IA 50021-9764
Tel: (515) 289-2331; web: www.swcs.org

UNIT 10

Waste Not,
Want Not

Bill Rathje, an archaeologist and "garbologist" at the
University of Arizona, studies the problem and the history
of solid-waste disposal in his excavation of a landfill in
Sunnyvale, California.

BEFORE YOU VIEW THE TELEVISION PROGRAM

Learning Objectives

After completing the assigned readings and viewing "Waste Not, Want Not," you should be able to

- describe, in general terms, the geological origins of mineral resources used by humans.

- list the environmental impacts associated with mining and explain how those impacts can be reduced.

- define depletion, and explain the economic and environmental implications of depletion of mineral resources needed by human society.

- discuss the origins and possible destinations of the materials in the following common items: the aluminum in a soft-drink can, the newsprint in today's *New York Times*, and the whole-wheat bread in the sandwich you had for lunch.

- list the three categories of waste materials discussed in the television program and give two examples of innovative methods developed to reduce or treat each category.

- list two major environmental problems associated with each waste category and contrast efforts to manage those problems in two countries shown in the television program.

- compare the advantages of waste reduction, reuse of materials, and recycling as methods of managing waste disposal problems.

- describe some of the economic and social obstacles that have impeded wide adoption of recycling and list the attributes of successful recycling programs.

- describe how you would go about reducing the wastes generated by your household, workplace, university, and community—and explain why.

Reading Assignment

Choose the material from either textbook as your reading assignment. Your instructor might assign additional readings as well.

Living in the Environment
Chapter 14, "Mineral and Soil Resources," Sections 14-3 through 14-5
Chapter 20, "Water Pollution," Sections 20-4 and 20-5
Chapter 22, "Solid and Hazardous Waste"

Environmental Science
Chapter 12, "Water Resources and Water Pollution," Sections 12-5 through 12-7
Chapter 13, "Mineral and Soil Resources," Sections 13-1 through 13-4
Chapter 14, "Solid and Hazardous Waste"

Unit Overview

The natural geological, chemical, and living processes of the earth support a constant flow of materials that are transformed and exchanged between land, atmosphere, and oceans. In contrast to these uninterrupted cycles of the geosphere and biosphere, the use of materials by human society tends to be *linear*: Raw materials are mined from the earth's crust or harvested from its forests and fields, refined, fabricated into products, used, and ultimately discarded. As populations grow and nations industrialize and trade with one another, the discards accumulate. The worldwide proliferation of landfills, toxic dumps, and contaminated land and water is evidence that this linear flow cannot continue.

This unit of *Race to Save the Planet* examines the mounting problem of waste management and the need for new approaches in the way human societies use materials. The assigned readings examine the sources of minerals in the earth's crust, discuss the environmental impacts of mining and the adequacy of mineral resource supplies, describe the problems of municipal solid waste and toxic waste production, and introduce the prevailing and prospective methods of handling and disposing of the wastes generated by human activities.

The accompanying television program, "Waste Not, Want Not," explores innovative solutions to the problem of waste. Individuals and institutions in countries that span the world's economic and cultural spectrum have discovered profitable opportunities to recover materials and energy from refuse, reduce the generation of toxic industrial wastes, and treat sewage with biological methods that enhance rather than diminish environmental quality. The program highlights the common elements of successful approaches and reviews progress by communities that have encouraged waste reduction and adopted "integrated waste management," a step toward replacing the linear mentality with a cyclic approach more consistent with the planet's natural metabolism.

Glossary of Key Terms and Concepts

The following terms and concepts will be helpful as background for viewing "Waste Not, Want Not."

Hazardous waste is a class of waste materials that poses immediate or long-term risks to human health or the environment and requires special handling for detoxification or safe disposal. Both industrial and household wastes include hazardous materials.

Incineration refers to a method of disposing of municipal solid waste and some hazardous chemical waste by burning it at high temperatures in specially designed combustion chambers.

Landfill is a site designated for disposal of solid or chemical wastes by burial. It may be essentially an open pit or a highly engineered facility that includes special linings to prevent wastes from leaking into water supplies.

Municipal waste includes all the solid waste generated by households and commonly collected on a community-wide basis by local government authorities.

Recycling encompasses a set of methods for recovering discarded materials and refashioning new materials of the same type; glass, paper, aluminum, steel, and some plastics are recycled commercially. Recycling generally involves an intermediate step in which the discarded material is melted down or processed before creating a new material, rather than reused directly.

Sewage treatment is a multistage process in which human wastes are filtered, decomposed, and purified to prevent the pollution of watercourses by nitrogen, phosphorus, and pathogens. Conventional treatment, carried out at large facilities that use a variety of mechanical and chemical means to decontaminate sewage, produces liquid effluent and solid sludge that is either used to fertilize land or dumped in landfills.

Waste-to-energy plants are solid-waste incinerators designed to produce steam or generate electricity that can be sold to local utilities as part of an urban power supply system. They are widely used in Europe and Japan and are increasing in popularity in the United States.

Zero-discharge technology describes a class of industrial processes designed to recover solvents, chemicals, and cleaning rinses used in manufacturing by collecting them and removing dissolved and suspended materials so that the liquids can be reused.

AFTER YOU VIEW THE TELEVISION PROGRAM

Consider What You Have Seen

"Waste Not, Want Not" takes an in-depth look at just one part of the materials system: the management, disposal, and reduction of waste. In many parts of the world, businesses, households, and communities have begun to reduce their waste, recover valuable materials from the waste stream for reuse, and pattern disposal practices on the planet's natural material cycles. The following themes relate the program to the reading assignment.

- Efficiency, policy, and economics
- Solid waste and municipal refuse
- Hazardous and industrial waste
- Coping with sewage

Efficiency, Policy, and Economics

Keep in mind the following three concepts that run through much of the new thinking on waste management.

Efficiency: Industrial processes based on recycled materials such as aluminum and paper are less costly and use less energy than processes using virgin materials; manufacturers that recover and reuse hazardous chemicals save on both production costs and waste disposal.

Policy: Stiffer enforcement of local regulations and national laws has made indiscriminate dumping of waste, especially toxic waste, more risky for dumpers, while heightened public awareness has cast an unfavorable spotlight on nations seeking to export their industrial waste to developing countries.

Economics: The rising cost of solid-waste *disposal* has prompted serious consideration of alternatives ranging from incineration to mandatory recycling in countries around the world.

Driven by necessity, waste management has become a field of startling vitality and innovation. Yet practices are by no means uniform around the world; even countries at similar levels of economic development reveal striking contrasts in their handling of wastes. The stories you have seen highlight two waste management gaps: the gaps between current practices in different societies and the gap between present disposal methods and the waste management practices envisioned by innovative researchers. The path to economic and environmental sustainability involves closing both gaps.

Solid Waste and Municipal Refuse

"Waste Not, Want Not" profiles a number of different communities and their response to the problem of disposal of waste.

- The program begins with landfills, places where most solid waste ends up. What are some of the things you would expect a landfill such as Fresh Kills on Staten Island to reveal about contemporary American life? Try to think of some aspects of contemporary life for which evidence couldn't be collected from landfills.

- In Lima, Peru, the television program shows that recycling is sometimes a means to bare economic survival, not a luxury that becomes possible only when societies become sufficiently affluent. What are your reactions to the scenes from Lima?

- Historically, recycling in the United States has been most widespread during times of hardship, such as World War II. But Joe Garbarino's successful community recycling program in northern California serves well-to-do neighborhoods. How do you account for this apparent paradox?

- What keeps Joe Garbarino in business?

- The tissue paper exchange, *chirigami kokan*, is a longstanding tradition in urban Japan. Do you think this kind of urban paper recycling could work in the United States? Why or why not?

- The recycling program in Machida, Japan, depends on a level of public participation that has few parallels in the United States. List some of the societal differences shown in "Waste Not, Want Not" that account for high participation in the Machida program.

- Describe the relationship of the incinerator to the recycling program in Machida. It is possible to look at the Machida program as an impressive but still preliminary demonstration of integrated waste management—or as an elaborate public relations campaign to gain community acceptance of an unwanted facility (the incinerator). What is your feeling about the Machida program?

- Waste incineration, widely opposed in the United States, consequently accounts for a very small share (about 3%) of the country's solid-waste disposal. By contrast, incineration is the rule in Japan, where an estimated 64% of solid waste is burned, although Japanese citizens also oppose new incinerators in their communities. Do you think the United States or Japan has a more appropriate level of dependence on waste incineration?

- What questions would you ask Mr. Tanaka, the chief of waste management research at Japan's Health and Welfare Ministry, about the Machida facility?

Hazardous and Industrial Waste

- The Kommunekemi facility in Nyborg, Denmark, shows the potential of partnership between government and private industry to improve the management of industrial waste. The Los Angeles County hazardous-waste-enforcement squad shows how a more adversarial relationship between government and industry can also result in control of industrial-waste dumping. One approach relies on good faith and common objectives, the other relies on good laws and effective enforcement. Does either approach offer an ideal solution to hazardous-waste management? Why or why not?

- What aspects of the Kommunekemi program could be transferred to other countries? What aspects do you think are unique to Denmark?

- The zero-discharge technology at Aeroscientific Corporation in California represents a new approach to the design of industrial processes. The program mentions several factors that have made zero discharge affordable at companies like Aeroscientific. List five of those factors. If you were the president of Aeroscientific Corporation, which factor or factors would you consider in deciding whether or not to adopt zero-discharge practices?

Coping with Sewage

- The contamination of water due to improper disposal of human wastes ranks among the world's most serious public health problems. Waterborne diseases spread by excrement account for millions of deaths in developing countries, including the preventable diarrhea that kills 5 million infants and young children each year. But the organic elements in human wastes—chiefly nitrogen and phosphorus—can potentially be recovered and put to productive use, just as wastes are automatically cycled in natural ecosystems.

- What phase of conventional sewage treatment does the Arcata, California, marsh system replace? What treatment functions does the marsh carry out?

- How do the sewage settling ponds (wastewater lagoons) in San Juan de Miraflores, Peru, differ from the waste treatment system in Arcata, California? Describe the function of the fish in this sewage treatment system.

- What other "products" does the San Juan de Miraflores waste treatment system provide? How does the system contrast with the concept of waste "disposal"?

Take a Closer Look at the Featured Countries

Denmark, a small West European country of 5.3 million people, is located on a peninsula that juts into the North Sea (Figure 10-1). With a population neither growing nor shrinking appreciably and an economy based largely on agriculture, Denmark enjoys high living standards and has a gross national product per person comparable to Japan's, roughly $34,890 per capita.

Not heavily industrialized, Denmark does not produce large quantities of hazardous or industrial waste. But virtually all of the country's drinking water comes from groundwater, which makes

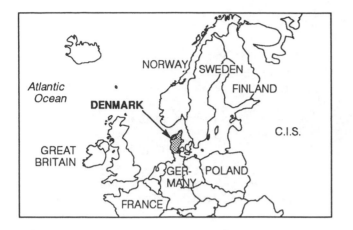

Figure 10-1   Denmark

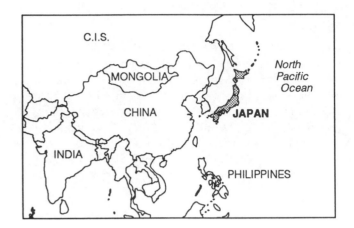

Figure 10-2   Japan

contamination by toxic chemicals an acute risk. According to the Organisation for Economic Co-operation and Development, an association of European governments, Denmark produces an estimated 814,000 metric tons of industrial waste each year, of which 63,000 tons are classified as hazardous and special wastes requiring treatment at the Kommunekemi facility. This modest amount of hazardous-waste generation, coupled with the country's affluence, accounts for the success of its waste treatment system.

Japan, a densely settled island nation, has a population of 127 million, roughly half that of the United States (Figure 10-2). Yet Japan's many islands together comprise an area no larger than the state of California. This limited land area and the rapid industrial development that has made Japan's economy second in size only to that of the United States make waste management an urgent national concern. Japan's economy, based on industrial exports, generates large quantities of industrial waste requiring management. The growing affluence of the heavily urbanized population contributes growing amounts of municipal refuse.

Japan's waste management strategy is based largely on resource recovery and incineration. Although Japanese are far more frugal than Americans, producing half as much refuse per person each year, the Japanese are second only to the Americans in the total quantity of municipal solid waste they must manage. The share of wastes recovered and reused or burned is far higher than in North America, and consequently Japan is much less reliant on landfills for disposal of either municipal refuse or industrial waste.

Peru, a mountainous country on South America's Pacific coast, is a poor country of 27 million people with a gross national product per person of $2,610 (Figure 10-3). The city of Lima, the country's capital, is home to more than a quarter of all Peruvians. The national population is growing at 2.2% each year, but Lima is growing even faster as rural families come to the capital city in search of jobs. Lima already has more residents than the country of Denmark. Because of its high birth rate, 35% of Peru's population is younger than 15 years.

Like most developing countries, Peru has no comprehensive policy for waste management. Human wastes in Lima's burgeoning neighbor-hoods pose a serious problem, because the density of the communities provides ideal conditions for the spread of wasteborne disease. According to United Nations estimates, only 57% of Peru's urban citizens have access to sanitation services of any kind. But major public investments in waste treatment plants or other expensive facilities are unlikely. Like its Latin American neighbors, Peru carries a heavy burden of foreign debt. Meeting the repayment obligations on this debt requires funds that the government could otherwise devote to the country's development.

The **United States**, with a population of 273 million, is in a class by itself in the production and disposal of waste materials (Figure 10-4). With the world's largest economy and highest income levels, the United States generates more waste material in total—and per person—than any other society. Japan produces about a third as much municipal solid waste as does the United States but dumps

Figure 10-3   Peru

Figure 10-4   The United States

less than one-eighth as much. U.S. communities and businesses produce an estimated 178 million metric tons of municipal solid waste, and 5.5 billion metric tons of potentially hazardous waste (with only 6% regulated by law), most of it contaminated water, each year. The garbage or municipal harzardous waste produced in the United States makes up only about 1.5% of the country's estimated total solid waste produced each year. The other 98.5% comes from mining, fossil fuel production, agriculture, and industrial activities.

Most solid and industrial waste still ends up in the land. Nine out of ten tons of U.S. hazardous waste, for example, are disposed of in landfills, impounded in settling ponds, or discharged to rivers and streams. Because of public opposition, incineration of both municipal waste and industrial toxic waste plays a smaller role in the United States than it does in most other industrial societies.

The United States incinerates about 17% of its solid waste, compared with 34% in the former West Germany, 51% in Sweden, and 64% in Japan. And the amount of useful materials recovered from the waste stream remains far below its potential. In 1992, the United States recycled about 30% of its aluminum, 39% of its paper, only 10% of its glass, and about 2.5% of its plastics. As landfill capacity is reached in more and more places, communities around the country are finding that they can no longer postpone consideration of new waste management approaches.

Examine Your Views and Values

1. What obstacles prevent widespread adoption of regional hazardous-waste-management approaches such as Denmark's Kommunekemi program in the United States? Should local communities or regional authorities have the final say on matters of waste disposal? Why?

2. Is everyone in the world entitled to a material standard of living comparable to that in the United States? If you answer "yes," consider the physical implications of the decision. If you answer "no," how do you justify the vast inequities in human well-being implied by your choice?

3. Most developing countries lack comprehensive programs for managing solid waste and toxic chemical waste. In 1988, reporters uncovered a number of instances in which industrial countries offered to pay developing countries to dump hazardous wastes (including toxic chemicals and incinerator ash) that could not be disposed of as cheaply or easily in their countries of origin. Consider the ethical issues raised by this international waste disposal. Who do you think should bear responsibility for the ultimate destination of any society's wastes? What factors should be weighed in making decisions about where to dispose of waste?

TEST YOUR COMPREHENSION

Self-Test Questions
(*Answers at end of Study Guide*)

Multiple Choice

1. Bill Rathje's studies of materials in American landfills have shown that
   a. most discarded materials break down into non-toxic components within a few years.
   b. plastics account for a growing share of American trash.
   c. there are fewer plastics in today's garbage than there were 20 years ago.
   d. packaging makes up the bulk of American garbage.
   e. More than one of the above is true. The correct answers are _____.

2. The Machida recycling center and incinerator show that
   a. Japan has patterned its waste disposal practices on the United States.
   b. Japanese citizens trust technology more than Americans do.
   c. advanced incinerators can control all air pollutants.
   d. advanced incinerators make recycling unnecessary.
   e. public attitudes are a key to acceptance of waste management facilities.

3. Aeroscientific Corporation installed its zero-discharge process in order to
   a. bring the company into compliance with federal water pollution laws.
   b. save money by recovering valuable materials used in manufacturing circuit boards.
   c. qualify for federal tax incentives awarded to efficient industries.
   d. respond to customers' demands for better pollution controls.
   e. catch up with standard practice in the electronics industry.

4. The sewage treatment ponds at San Juan de Miraflores, Peru, are
   a. a technology too sophisticated for most developing countries.
   b. a technology too primitive for the United States or other industrial countries.
   c. a cost-effective method of controlling waterborne diseases.
   d. a cost-effective method of providing safe drinking water.
   e. a short-term solution to the problem of sewage disposal.

5. The wastewater treatment lagoons in Arcata, California, show that
   a. biological treatment of human waste can replace conventional sewage treatment.
   b. biological treatment of human waste can complement conventional sewage treatment.
   c. most wild birds and animals will stay away from a wastewater lagoon.
   d. sewage can be released into coastal waters without lasting damage.
   e. biological treatment has to be done in remote locations.

6. Denmark's hazardous-waste-management program
   a. was largely patterned on U.S. practices.
   b. reduces waste volume by three-quarters.
   c. is run by the Danish army.
   d. works because Denmark is a highly regimented society.

7. Raw materials from biological sources
   a. are always renewable because living things reproduce themselves.
   b. cause little environmental damage because they come from nature.
   c. are used in the same way in industrial and developing countries.
   d. are renewable in principle but can be depleted in practice.

8. Most hazardous chemical wastes in the United States end up
    a. in landfills lined to prevent leakage into water supplies.
    b. in landfills, surface lagoons, and injection wells.
    c. reused by the industrial processes that generate them.
    d. burned in high-technology incinerators.

9. Conventional sewage treatment
    a. produces nothing except an outflow of clean water.
    b. generates sludge that must also be disposed.
    c. generates no valuable by-products.
    d. is a relatively inexpensive way to dispose of human waste.

True or False

1. Recycling and incineration together account for more than half the solid waste disposed of in the United States. _____

2. Japan incinerates about 3% of its solid waste. _____

3. The United States incinerates about 3% of its solid waste. _____

4. Paper recycling is more widespread in the United States than in Japan. _____

5. Japan's limited size makes landfills a disposal method of last resort. _____

Sample Essay Questions

1. List and explain three ways in which waste disposal strategies are linked to the greenhouse effect.

2. You are town manager of a community whose landfill is nearing capacity. Outline a presentation to the town council in which you explain the problem, review the community's options, and summarize the costs and benefits of each choice.

3. Should industrial nations like the United States and Japan be concerned with waste management practices in nations like Peru? Why or why not? Design an aid program for waste management in a developing country, using examples from the television program if appropriate.

4. Is waste management an important concern in development? Why or why not? Use examples from any of the television programs and from your readings.

5. Explain what happens to your own household's trash. How much do you think your waste stream could be reduced? In your answer, explain what happens to glass, newspaper, aluminum, plastics, and food scraps in your trash, and describe how waste management in your home could be improved.

GET INVOLVED

References

Dunning, Alan T. *How Much Is Enough: The Consumer Society and the Future of the Earth.* New York: W.W. Norton, 1992.

Frosch, Robert A., and Nicholas E. Gallopoulos. "Strategies for Manufacturing." *Scientific American*, Sept. 1989.

National Academy of Sciences. *The Greening of Industrial Ecosystems.* Washington, D.C.: National Academy of Engineering, 1994.

Rathje, William, and Cullen Murphy. *Rubbish!: The Archaeology of Garbage.* San Francisco: HarperCollins, 1992.

World Commission on Environment and Development. "Industry: Producing More With Less." *Our Common Future.* Oxford: Oxford University Press, 1987.

Young, John E. *Discarding the Throwaway Society*. Washington, D.C.: Worldwatch Institute, 1991.

Young, John E., and Aaron Sachs. *The Next Efficiency Revolution: Creating a Sustainable Materials Revolution*. Washington, D.C.: Worldwatch Institute, 1994.

The following chapters in *Taking Sides: Clashing Views on Controversial Environmental Issues*, relate to waste management.

- Issue 13. "Hazardous Waste: Can Present Regulatory and Voluntary Efforts Solve Disposal and Clean-Up Problems?" (pp. 216–231)

- Issue 14. "Municipal Waste: Is Modern Waste Incineration Technology an Environmentally Benign Alternative to Garbage Dumps?" (pp. 232–247)

You may find the following chapters on waste management and recycling issues from Worldwatch Institute's annual *State of the World* reports especially helpful.

- Anne Platt McGinn, "Phasing Out Persistent Organic Pollutants," *State of the World 2000*.

- Gary Gardner and Payal Sampat, "Forging a Sustainable Materials Economy," *State of the World 1999*.

- Gary Gardner, "Recycling Organic Wastes," *State of the World 1998*.

- Hal Kane, "Shifting to Sustainable Industries," *State of the World 1996*.

- John E. Young, "Reducing Waste, Saving Materials," *State of the World 1991*.

- Alan Durning, "Asking How Much Is Enough," *State of the World 1991*.

- Sandra Postel, "Controlling Toxic Chemicals," *State of the World 1988*.

- Cynthia Pollock Shea, "Realizing Recycling's Potential," *State of the World 1987*.

- William U. Chandler, "Recycling Materials," *State of the World 1984*.

Organizations

Many communities in the United States have voluntary recycling programs that welcome newspapers, glass containers, and aluminum cans, and mandatory programs are becoming common as landfills reach capacity. If you don't already recycle, a call to your local government, or to a nearby college or university, should be all you need to locate a recycling program in your area.

Both local and national environmental organizations are active in recycling and waste reduction activities. The two national groups listed below support recycling education and could provide additional information and the names of groups involved with recycling in your area.

Center for Health , Environment, and Justice
P.O. Box 6806
Falls Church, VA 22040
Tel: (703) 237-2249; web: www.essential.org/cchw

Environmental Defense Fund
257 Park Ave. South
New York, NY 10010
Tel: (212) 505-2100; web: www.edf.org

PART III

STEPS TOWARD
SUSTAINABILITY

UNIT 11

It Needs Political Decisions

April 1990 marked the twentieth anniversary of Earth Day, a demonstration calling for environmental awareness and activism around the world.

BEFORE YOU VIEW THE TELEVISION PROGRAM

Learning Objectives

After completing the assigned readings and viewing "It Needs Political Decisions," you should be able to

- explain why governments and their citizens sometimes find themselves at odds over protection of the environment and discuss how clashing interests have been resolved in each of the three featured countries.

- compare the methods and accomplishments of national efforts to slow population growth in Zimbabwe and Thailand.

- describe the causes and consequences of land degradation in each of the three featured countries and explain the role of each government in addressing land degradation problems.

- describe and evaluate Thailand's efforts to address the environmental impacts of industrial development.

- explain how Sweden has begun to address pollution by encouraging industries to reformulate their products.

- discuss how education has helped Zimbabwe, Thailand, and Sweden pursue national efforts to address environmental problems.

- explain how citizen pressure can change government environmental policies.

- use an example from "It Needs Political Decisions" to show how a decision to protect the environment *within* national borders can sometimes cause environmental damage *beyond* national borders.

Reading Assignment

Choose the material from either textbook as your reading assignment. Your instructor might assign additional readings as well.

Living in the Environment

Chapter 27, "Economics and Environment"

Chapter 28, "Politics and Environment"

You may wish to review Sections 11-3 and 11-4 of Chapter 11, "Human Population"; Chapter 14, "Minerals and Soil Resources," Sections 14-6 and 14-7; Chapter 23, "Sustaining Ecosystems," Sections 23-2; and Chapter 24, "Sustaining Ecosystems: Deforestation, Biodiversity, and Forest Management."

Environmental Science

Chapter 2, "Economics, Ethics, and Sustainability," Sections 2-1 through 2-6

You may wish to review Chapter 9, "The Human Population," Sections 9-3 and 9-4; Chapter 13, "Mineral and Soil Resources," Sections 13-6 through 13-8; and Chapter 17, "Sustaining Terrestrial Ecosystems: Forests, Rangelands, Parks, and Wilderness," Sections 17-1 through 17-6.

Unit Overview

This unit of *Race to Save the Planet* looks at three nations that have taken steps to tackle environmental problems and to pursue sustainable development at the national level. The process of national decision making about the environment involves public education, difficult trade-offs that force governments to resolve competing economic interests, and participation by private citizens in matters of public policy. Governments, whether they are elected or rule by force, claim concern for the quality of life of their citizens. Today it is becoming clear to governments around the world that management and protection of the environment is not just another item among the many that crowd their agendas but the foundation for progress on improving quality of life.

"It Needs Political Decisions" reviews national experiences in devising environmental strategies in Zimbabwe, Thailand, and Sweden. These countries span the economic and demographic spectrum, from Zimbabwe, where rural poverty is widespread and the population is expanding by 3.6% each year, to Sweden, which is among the world's most affluent societies and has nearly attained zero population growth. Despite enormous differences in standards of living, these three societies face common problems of land and resource management and the environmental consequences of national efforts to promote economic development. In each, the government has played a major role in setting the environmental agenda, but each has taken different steps to reconcile the government's concerns with those of its citizens.

The assigned readings in *Living in the Environment* and *Environmental Science* review the political and economic dimensions of environmental decision making, with an emphasis on the United States. Whichever text you use, take time to review the chapters that deal with population control, soil resources, and forestry before watching "It Needs Political Decisions." These are among the substantive issues that must be dealt with by every country that chooses to address environmental problems at the national level.

Environmental politics is shaped by the answers to several key questions: What are the government's interests? What are the people's interests? Where do those interests clash? Where do those interests converge? Who leads, and who follows, in the effort to devise responses to environmental problems? Ultimately, these questions of citizenship and authority must be faced by all societies, as the concept of "planetary citizenship" that is part of the environmental revolution gains recognition around the world.

Glossary of Key Terms and Concepts

The following terms are introduced in "It Needs Political Decisions." Review them once now, before viewing the program.

Colonialism was a system in which several European nations governed developing countries in Africa, Asia, and Latin America in order to exploit their mineral and natural resources and low-cost labor. Most former colonies became independent nations in the 1960s and 1970s, some only after bitter wars of independence.

Zimbabwe's **communal lands**, areas set aside for settlement and farming by the country's black majority, were created by a preindependence government that reserved the country's best lands for white settlers. The communal lands include the country's least fertile areas, characterized by poor soils and low and erratic rainfall.

The **Gothenberg project** is a local effort to make the heavily industrialized city of Gothenberg (Göteborg), Sweden, a national leader in environmental quality. One innovative part of the project involves collaboration with manufacturers based in the city to replace environmentally damaging products with alternatives that are benign and nonpolluting.

The network of **klongs**, or canals, that run through Thailand's capital, Bangkok, have led some to call the city the "Venice of the East." Today the klongs are heavily polluted by untreated sewage and wastes from industries located in the city.

Over 38 countries have completed **national conservation strategies** in which environmental priorities and opportunities for sustainable management of natural resources are highlighted, following the example of the World Conservation Strategy published by the International Union for the Conservation of Nature and Natural Resources (IUCN) in 1980. Though governments may support preparation of the strategies, they are not bound to follow the IUCN's recommendations.

The **newly industrializing countries** include several Southeast Asian nations (South Korea, Taiwan, Thailand, Malaysia) that have achieved high rates of economic growth in recent years by attracting manufacturing and assembly plants for the automotive, electronics, and other industries. The industries have benefited from relatively educated workers, low wage levels, various sorts of government incentives, and lax environmental regulation.

Plantation forests such as those in Sweden are managed stands of commercial timber trees, usually planted to a single species and fertilized and "weeded" like cropland. Tree species grown in plantations are selected for their rapid growth and typically harvested by mechanized clearcutting.

The bite of the **tsetse fly**, common in parts of central and southern Africa, can transmit a parasite that causes trypanosomiasis, or sleeping sickness, a potentially fatal disease of cattle and humans. Prevalence of the tsetse has prevented settlement of areas such as Zimbabwe's Zambezi Valley, shown in the television program.

AFTER YOU VIEW THE TELEVISION PROGRAM

Consider What You Have Seen

"It Needs Political Decisions" profiles three very different nations that have used the tools of government and politics to devise ambitious responses to environmental problems. Experience

in Zimbabwe, Thailand, and Sweden shows that government policy is a necessary but not sufficient condition for achieving sustainable and environmentally sound development. Another key dimension is the role of individual citizens, whose choices and behaviors can in some cases thwart government goals and in other cases prod reluctant governments into action. As you reconsider the television program, keep in mind this balance between the centralized authority of the state and the responsibility of the individual.

Use the following themes to organize your review of the parallels and contrasts between the three featured societies.

- Population

- Land use

- Industrialization and pollution

- Decision making: Striking the balance

Population

Take a few minutes to think about the population dynamics in the three societies shown in the program. Table 11-1 summarizes the basic population statistics.

TABLE 11-1   BASIC POPULATION STATISTICS

Country	1999 Population (in millions)	Year 2025 Population (in millions)	Growth Rate (percent)	TFR[1]	Percent of Married Women Using Contraception
Zimbabwe	11.2	12.4	1.2	4.0	48
Thailand	61.8	73.0	1.1	2.0	72
Sweden	8.9	9.3	- 0.1	1.5	78

[1]TFR = Total fertility rate, or average number of children that would be born to each woman during her childbearing years assuming current birth rates remain constant; a TFR less than 2.1 is considered below replacement level.

SOURCE: Population Reference Bureau, *1999 World Population Data Sheet* (Washington, D.C.: 1999).

Compare the effect of total population size on expected growth by the year 2000 in the three societies. Zimbabwe, with the fastest growing population, will add about 1.2 million people (equal to its present population) over the next 26 years. Thailand, though growing slower than Zimbabwe, will add about 11 million from its much larger base. Meanwhile, Sweden's population will only grow slightly in the next 26 years. Why is that, when the figures show it to be still growing by a 0.1% per year? The answer lies in the total fertility rate (TFR), or the average number of children each woman will bear during her childbearing years. Swedish mothers bear, on average, fewer children than needed to replace them in the next generation. There are still enough Swedish mothers that the population is growing very slightly. But by the turn of the century, the population will have begun to shrink. Sweden is said to be at zero population growth.

Why are governments concerned about the rate of population growth in their societies? In Zimbabwe, the government's primary concern is the country's pervasive poverty. Rapid growth means that the government can do little more than run in place in its effort to provide basic services. Equally important in a country whose people depend on subsistence farming is the balance between people and land. As a family-planning worker shown in the program says, "The land is not growing. It's us who are growing." In Thailand, the government sees slower population growth in terms of avoided births and calculates the savings in services and expenditures that can be devoted to better meeting the needs of the existing population. In the television program, Mechai Viravaidya, the designer of the Thai family-planning program, expresses an even more fundamental political concern: More people whose needs cannot adequately be met ultimately means a risk of discontentment and political instability—which can put the government itself in jeopardy. Part of the Thai government's concern is self-preservation. Sweden has "no population problem," but the government sees the quality of life as its top priority. Even at zero population growth, the Swedish government supports family-planning services and child care, so that the benefits of the country's affluence can be shared widely in the society.

Compare the government's approach to population in Zimbabwe and Thailand. Why have both governments chosen to emphasize education in the effort to combat high growth rates? What is the basic message about small families that each country tries to communicate? What evidence does the television program offer to support the idea that small families can have a better life? Thailand's approach involves not only making contraceptives available so that people have the means to plan their families but also making contraceptives familiar and acceptable. This approach, known as "social marketing," has been used in many societies.

Land Use

Land management and land degradation pose some of the most difficult challenges on government agendas in the countries shown in "It Needs Political Decisions." Each government has had to make decisions about which land should be cultivated or harvested and which land should remain wild, and to strike a difficult balance between competing or incompatible land uses. In Zimbabwe, the fundamental issue is food production—how to meet the needs of a growing population for food without encouraging settlement of inappropriate land or jeopardizing wildlife areas that attract tourists and bring the country much-needed foreign exchange. In Thailand and Sweden, the most pressing issue are that of forest management. Timber from Thailand's forests has long been one of the country's most valued exports, but the dwindling forests that remain also protect densely settled lowland rice-growing areas from catastrophic floods. Sweden's still-abundant forests, the base of the country's affluence, have been turned into uniform commercial plantations scarcely resembling the native forests that once blanketed the country; some Swedes wonder whether single-minded pursuit of commercial productivity has left their forests vulnerable to catastrophe and sacrificed too easily the many other benefits of wild ecosystems.

Each of the three countries has faced inescapable trade-offs in a characteristic way, and sometimes the government has had a hard time deciding what the best policy is. On the one hand, Zimbabwe's government supports efforts to eradicate the tsetse fly, which will make the valley suitable for farming and grazing. In a somewhat conflicting move, the government is also developing an innovative approach to wildlife management that will give local people a financial stake in the region's unique fauna. The plan is an uncertain gamble that agriculture and wildlife conservation can coexist, but Zimbabwe's government believes that unless people benefit from wildlife, farms and cattle will inevitably displace it. Do you think that Zimbabwe is justified in this effort to, in effect, put a price-tag on its wildlife? Is there another way to ensure that animals and people coexist?

The floods of late 1988 forced the government of Thailand to face a difficult trade-off: How could it protect Thai citizens from destructive floods without in turn restricting the politically powerful and economically important wood products industry? Thailand made a choice that was on the one hand courageous and on the other politically expedient: The Thai prime minister declared a ban on logging *within* Thailand, and the country began to import timber from its neighbors (some of whom cut trees within the same upland watersheds where the Thai floods originated) to keep its lumber mills and wood export businesses operating. In a sense, Thai authorities made a decision to protect Thailand at the expense of her neighbors, whose short-term economic gains outweigh concern for forest protection. What other choices should Thai authorities have considered?

Sweden once had a deforestation crisis and responded by passing a national law that required cut forests to be replanted. As a result, Sweden today has more trees—nearly 60% of its land territory—than do most nations but ironically has less *natural* forest than does either Thailand or Zimbabwe. Sweden has substituted fast-growing, easily harvested commercial forests for the diverse native boreal forest that once cloaked the country. What are some of the reasons that biologists are concerned about the loss of the native forests? As in Thailand, the forest products industry in Sweden is economically important and therefore politically influential. The decision to sacrifice natural diversity for the sake of economic productivity is one that has been made to varying degrees by all societies. How does Sweden's choice compare with the trade-offs made in Zimbabwe and Thailand?

Industrialization and Pollution

Industrial pollution, perhaps the most familiar challenge to governments on matters of environment, has not yet been "solved" by any society. Thailand, a newly industrializing country, is just beginning to build the expertise and create the institutions needed for effective pollution control, whereas Sweden has environmental laws, trained personnel, and industries as advanced as any country on earth. Yet both countries struggle to achieve desired levels of air and water quality, with Thailand just beginning to impose *output* controls on polluters while Sweden, under pressure from its citizens, is moving toward *input* controls and designing nonpolluting products and equipment. Again, progress involves resolving tensions between environmental quality and influential economic interests.

The problem of pollution is inescapable in Thailand's crowded capital, Bangkok. The growing number of automobiles brings pollution and smog problems like those in Los Angeles, discussed in Unit 5. Industries and households dump wastes directly into the network of canals that laces through Bangkok. Problems of contamination have grown far more rapidly than the government's ability to respond.

First alerted to pollution problems, governments typically respond by creating pollution control authorities and setting standards for air and water quality. Thailand did this by establishing the National Environment Board in the early 1970s. Unfortunately, the board has never had financial resources sufficient to carry out its mission. Though it has some limited enforcement responsibilities, the board's primary role is public education to spread awareness of the need for pollution controls. The small manufacturing and industrial shops largely responsible for Thailand's robust economic growth in recent years are also the prime polluters of air and water. Not surprisingly, the Thai government is reluctant to impose pollution control requirements that would threaten the country's new-found prosperity; it hopes that education will achieve the goals it has so far been unwilling to pursue through vigorous regulation and enforcement.

Even in Sweden, industries will pollute until pollution control laws are enforced, and the government is constantly challenged by citizen activists to enforce output controls more energetically. Public pressure like that in Sweden has been important in most countries that have responded to conventional pollution of air and water. Sweden has recently gone further, though, by using government authority to challenge manufacturers to reformulate products that cause environmental damage. "It Needs Political Decisions" presents two examples, both part of the imaginative Gothenberg project: the soap manufacturers whose product was linked to toxic sediments that had accumulated in Gothenberg Harbor, and the pipe maker required to find a substitute for chlorofluorocarbons (CFCs) in the manufacture of insulated pipes.

This latter example in particular is an important one, because it shows that economic concerns do not *always* prevail in government environmental decision making. Even with government support to develop the alternatives, CFC-free pipes proved to be 10% more expensive than earlier pipes—enough of a margin to hurt the company's sales. The government policy reflected a decision that, in this case, the quality of the environment and the global atmosphere took precedence over the economic interests of the manufacturer. Policies like these help move industries in the direction of sustainability, sooner than purely economic factors would lead them to do so.

Why do you think authorities in Bangkok might be reluctant to tackle pollution the way authorities in Gothenberg did? Think of some of the ways that government in both countries has acted as an obstacle to effective pollution control, and of ways that each government has tried to encourage innovation. Which approaches seem to you most likely to be transferable to other societies?

Decision Making: Striking the Balance

In all the examples shown in "It Needs Political Decisions," government authorities seem never to have an entirely clear mandate to protect environmental quality. There are always countervailing considerations, usually having to do with short-term economic benefits. Environmental decision making at the national level is always a matter of finding the balance. Where that balance lies is

sometimes shaped by the way authority is shared between governments and citizens within a society.

The three featured countries provide some striking contrasts on this point. In Zimbabwe, the government leads, but listens, on matters of environmental policy. Its greatest challenge—population growth—is not a problem that forceful coercion is likely to solve. Government efforts are instead devoted to convincing people that having smaller families will bring them a better life. Thailand proves that this approach can succeed in family planning, but education may not be enough to control pollution or restrain indiscriminate deforestation. Governments often evade tough decisions on these issues, which involve vested economic interests, unless public opinion forces them to do so. In Thailand, which has no long tradition of citizen activism or public participation in politics, that popular voice has only begun to emerge in recent years. In Sweden, public pressure can shape national policy, and citizens push their government to go further than current law requires, to adopt new policies and regulations prompted by public concern. This degree of public participation, common in industrial democracies, is only slowly spreading to less developed societies.

Review the three case studies, and think of the different ways the three governments seek to influence their citizens and affect individual behavior. Then consider how each government treats established industries or economic concerns. How do governments set examples for their citizens? How can citizens set an example for their government? After watching this program, how important do you think popular participation is in making real progress on environmental problems?

Take a Closer Look at the Featured Countries

The Kingdom of **Sweden** is a modern European state still ruled by a constitutional monarchy whose roots go back more than 1,000 years (Figure 11-1). Like its Scandinavian neighbors, Sweden is a progressive industrial society whose citizens equitably share the country's affluence. With a slightly declining population of 8.9 million, the country's gross national product per person was

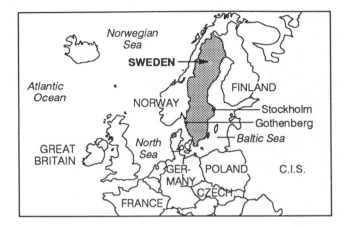

Figure 11-1   Sweden

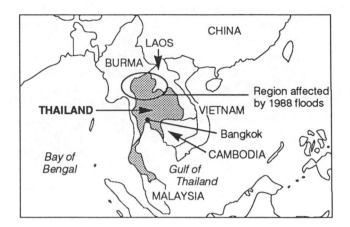

Figure 11-2   Thailand

$26,210 in 1997 (compared with $29,080 in the United States). Sweden's infant mortality rate, a key social indicator, was 3.6 per 1,000 live births, one of the lowest rates any society has achieved. By most measures, the quality of life of Swedish citizens is extraordinarily high.

Though the nominal head of the government of Sweden is King Carl Gustaf, the day-to-day affairs of the country are governed by a prime minister and a legislature called the Riksdag whose members are elected by the people for three-year terms. In recent years, over 90% of the Riksdag members have been women. A lively array of political parties is represented in the Riksdag, reflecting a pluralistic and politically active populace.

Sweden has a longstanding commitment to neutrality in international affairs and a comprehensive welfare state that many other nations have sought to emulate: Sweden instituted old-age pensions in 1911 and uses government funds to support comprehensive health care, housing, and employment-security programs. Environmental concerns are in the forefront of Swedish politics. In a 1980 referendum, Swedes voted to limit the country's reliance on nuclear power and phase to out the country's reactors by 2010. The country's vulnerability to acid deposition and damage to lakes and forests heightened environmental awareness and aroused public interest in problems that stretch far beyond Sweden's borders.

The Kingdom of **Thailand** is, like Sweden, a nominal monarchy that is today ruled by a parliamentary democracy (Figure 11-2). Sand-

wiched between Burma, Laos, and Cambodia, Thailand has become a major economic power in Southeast Asia. By standards of other developing countries, Thailand is comparatively well off, with a population growing considerably more slowly than that of its neighbors, a dynamic economy, and a gross national product per person of $2,740 in 1997. The country's infant mortality rate, 25 per 1,000 live births, is nearly seven times higher than Sweden's but still among the lowest in Southeast Asia. More than 6 million people live in the densely settled and cosmopolitan capital of Bangkok. But seven out of ten Thai citizens live in the country's rural areas, dependent on rice cultivation.

Though Thailand's monarchy is today mostly symbolic, the king and royal family are important figures to most Thai citizens. Although much of Southeast Asia was controlled by Britain or France in the nineteenth century, an agreement between these countries in 1896 guaranteed Thai independence and self-rule. Representative government was first introduced in 1932. Ruling authority was held by the military rather than elected leaders until the mid-1980s. The country's present prime minister, Chatichai Choonhavan, the first civilian leader since 1976, was elected in 1988.

With a still-young tradition of participatory democracy, citizen activism is just beginning in Thailand. But environmental concerns in particular are helping to change that. Public outcry following the catastrophic floods of late 1988 helped push the prime minister to declare the first-ever logging ban in Thailand, over the objections of powerful logging interests. Demonstrations and citizen pressure in

Bangkok advocate a stronger government commitment to environmental quality. Whether damage to Thailand's environment, particularly the overuse of the country's forests, has already gone too far to be corrected by enlightened government policy remains to be seen.

The Republic of **Zimbabwe**, a landlocked southern African country encircled by Botswana, Zambia, Mozambique, and South Africa, emerged only recently from a long colonial history of minority white rule (Figure 11-3). Settled originally by a number of distinct tribal groups, Zimbabwe was colonized by the British South Africa Company at the end of the nineteenth century and known as Rhodesia. The white-ruled government declared unilateral independence from Britain in 1965, and the next 15 years were disrupted by protracted guerrilla warfare as the black majority challenged white rule and the continuing division of the country along racial lines. A constitution based on black majority rule was adopted in 1979 and was formally recognized by Britain in 1980. The present government is led by Executive President Robert Mugabe.

Today Zimbabwe is gradually recovering from the legacy of civil war and coping with the demographic and economic problems common throughout Sub-Saharan Africa. The present population of more than 11.2 million is growing by 1.2% each year. The high fertility levels and rapid growth mean that the country's population is skewed toward the young, with 44% of all Zimbabweans under 15 years old. The infant mortality rate is 53 per 1,000 live births, considerably lower than in many African nations. Its estimated gross national product per person was $720 in 1997.

Although it is a republic based on majority rule, Zimbabwe does not yet possess a truly open political system. By consolidating various factions that had opposed white rule during the war years, Executive President Robert Mugabe effectively created a one-party state in late 1987. Thus, unlike Sweden's multiparty system, opposing views have no formal representation in Zimbabwe. But to its credit, Mugabe's government has worked to address the country's most pressing need, by creating a comprehensive and well-run program for

Figure 11-3   Zimbabwe

family planning and maternal and child health. In addition, the country has launched innovative efforts to restore and manage the environment by allowing local people to benefit economically from woodland and wildlife resources. A great remaining challenge is the more equitable distribution among the black majority of the country's most fertile land, which was reserved for whites only during the colonial era. As Zimbabwe makes progress on population, food production, and environment, its citizens may demand a stronger voice in the government.

Examine Your Views and Values

1. Do you think governments should set goals for national population growth or desirable population size? Why or why not? Should the United States have such a goal?

2. Bjorn Gillberg, the Swedish environmental activist shown in "It Needs Political Decisions," comments that "government agencies are nothing without people chasing after them." Do you agree or disagree with Gillberg? Have you ever "chased" a government agency to obtain a neglected service or right a wrong?

3. Choose one of the three countries shown in "It Needs Political Decisions." Write down your impressions after watching the program and compare them with the assumptions about that country that you might have expressed before watching. What had you expected to see that was contradicted by the program? What had you expected that was reinforced by the program? What specific scenes or sequences gave you different insight into the country you chose?

TEST YOUR COMPREHENSION

Self-Test Questions
(Answers at end of Study Guide)

Multiple Choice

1. The campaign to slow population growth is considered critical to Zimbabwe's future because
   a. the country's birth rate is one of the highest in the world.
   b. other problems of land use and pollution have been handled successfully.
   c. the present population is near the limits of the country's productive land.
   d. the government is anxious to keep people out of existing wildlife areas.
   e. More than one of the above is true.
      The correct answers are _____.

2. Which of the following is *not* among the elements of Zimbabwe's family planning program?
   a. education on basic health care for mothers
   b. supplies of contraceptives
   c. family-planning workers in every village
   d. direct payments to families who have no more than two children
   e. education on the benefits of smaller families

3. The government's plan for the Zambezi Valley involves
   a. construction of dams that will turn the valley into an industrial center.
   b. allowing some agricultural settlement and encouraging commercial activities based on wildlife.
   c. allowing some agricultural settlement and some industrial development.
   d. managing the valley's wildlife so that there is enough for subsistence hunters.
   e. relocating families from the crowded communal lands to the valley.

4. The government is anxious to preserve the Zambezi Valley's wildlife because
   a. wildlife is the best indicator of environmental quality in the region.
   b. wildlife is a prime tourist attraction that brings money into the country.
   c. environmentalists worldwide have put pressure on Zimbabwe.
   d. wildlife is sacred to most people in Zimbabwe.
   e. tsetse flies make the valley useless for most other purposes.

5. The rate of population growth in Thailand is
   a. about twice as high as the rate in Zimbabwe.
   b. now nearly the same rate as in Sweden.
   c. increasing to the same level as in Zimbabwe.
   d. about half as high as the rate in Zimbabwe.
   e. considered too low by the Thai government.

6. The main message of Thailand's national family-planning campaign is
   a. families with more than two children will be punished.
   b. condoms are the only acceptable method of birth control.
   c. life can be better for families that limit their number of children.
   d. using contraceptives is unethical.
   e. families with few children will become wealthy.

7. Thailand's National Environment Board carries out its mission by
   a. thoroughly enforcing regulations, with stiff penalties imposed on polluters.
   b. controlling automobile emissions at the expense of other pollutants.
   c. using education to supplement limited enforcement activities.
   d. using the royal family to convince people not to pollute.
   e. using high-tech monitoring equipment that reveals low-level pollution.

8. The shrinkage of Thailand's forest cover is mainly due to
   a. timber companies that ignored replanting rules.
   b. a growing population's need for farmland.
   c. government emphasis on agricultural development at the expense of forests.
   d. the eradication of tsetse flies that once made forests uninhabitable.
   e. More than one of the above is true. The correct answers are _____.

9. The king of Thailand and his family have been most concerned with
   a. providing rural communities with alternatives to forest destruction.
   b. developing new food crops and fighting malnutrition.
   c. promoting industrial development to modernize the country.
   d. relieving the deadly air pollution in Bangkok.
   e. winning back power from the Parliament and prime minister.

10. The role of environmental activists in Thailand is
    a. ignored by the king and the Thai government.
    b. much more limited than in Zimbabwe.
    c. beginning to focus government attention on long-neglected problems.
    d. traditionally a major source of new government policies.
    e. comparable to the role such activists play in Sweden.

11. In contrast to Zimbabwe and Thailand, Sweden has
    a. a population of only about 10 million.
    b. a population that is nearly stable.
    c. virtually no serious environmental problems.
    d. almost none of its original native forests left.
    e. More than one of the above is true. The correct answers are _____.

12. The survey of Gothenberg Harbor showed that many toxic chemicals in the harbor sediments came from
    a. ordinary detergents and other household products.
    b. waste from a detergent-manufacturing plant.
    c. oil refineries and other heavy industries in the city.
    d. sewage from households in the city.
    e. ships that rinsed their holds in the harbor.

13. Swedish environmentalists are concerned about the country's forest practices because
    a. forest practices involve clearcutting, which exposes large areas to erosion.
    b. nearly all the country's forests have been cleared for farmland.
    c. nearly all the country's natural forests have been converted to single-species plantings.
    d. the timber industry regularly ignores replanting requirements.
    e. More than one of the above is true. The correct answers are _____.

Sample Essay Questions

1. Why do governments and their citizens sometimes have very different interests in environmental protection? Discuss an example from "It Needs Political Decisions" that shows how individual needs and national needs can be at odds.

2.   Contrast the approaches to pollution control used in Sweden and Thailand.   Which approach is more likely to be successful in the long run?

3.   Do you think Thailand's decision to ban logging after disastrous floods struck in 1988 was a wise decision? Explain your answer.

4.   Why are there few environmental activists in Zimbabwe? What do you think this means for government environmental policy there?

5.   Sweden's national forestry policy encourages more reforestation than most countries ever accomplish. In what ways is this policy environmentally questionable?

GET INVOLVED

References

The following publications deal with what nations can do to tackle environmental problems.   The easiest to obtain are the book by Timberlake and the report by the World Commission on Environment and Development, which you should be able to find in your library.

International Union for the Conservation of Nature and Natural Resources (IUCN). *World Conservation Strategy*. Gland, Switzerland: IUCN, 1980.

IUCN, the World Wide Fund for Nature, and the United Nations Environment Programme. *Managing for the Future: The Second World Conservation Strategy Project*. Gland, Switzerland: IUCN, available in 1991.

Timberlake, Lloyd. *Only One Earth: Living for the Future*. New York: Sterling Books, 1987.

World Commission on Environment and Development. *Our Common Future*. New York: Oxford University Press, 1987.

Organizations

Zimbabwe is just one of more than 30 developing and industrial nations around the world that have completed national conservation strategies. These strategies are developed by governmental and nongovernmental organizations, often with the support of the International Union for the Conservation of Nature and Natural Resources (IUCN).

Though it does not always result in a published document, the process of assembling a national strategy gives authorities a chance to integrate conservation and environmental concerns into national development plans. Most strategies are patterned on the world conservation strategy developed in 1980 by the IUCN. Information on many of the national strategies can be obtained from the IUCN's headquarters in Europe. A new edition of the *World Conservation Strategy* is being prepared and will be published before a United Nations Conference on the Environment scheduled for 1992.

IUCN: The World Conservation Union
128 Rue Mauverney
CH-1196 Gland
Switzerland

Now or Never

A man stands on a soot-laden snowbank amid the iron and steel mills of Sverdlovsk. In the Commonwealth of Independent States, an estimated 102 cities have air pollution ten times the allowable level.

SOVFOTO

BEFORE YOU VIEW THE TELEVISION PROGRAM

Learning Objectives

After completing the assigned readings and viewing "Now or Never" you should be able to

- describe Kenya's greenbelt movement and explain the factors responsible for its success. discuss two examples of environmental deterioration in the former Soviet Union that have provoked a public response.

- explain the relationship between environmental concerns and the political policy known as *glasnost* (openness) in the former Soviet Union.

- describe the Mediterranean Action Plan and discuss its significance for other international efforts to manage shared resources.

- compare and contrast the perspectives of industrial countries and developing countries concerning international efforts to phase out production of ozone-damaging chlorofluorocarbons (CFCs).

- explain the significance of the Montreal Ozone Agreement for subsequent international agreements to address problems of the global atmosphere.

- discuss the relationship between reduction of East–West political tensions and progress on global environmental concerns.

Note: The political boundaries of eastern Europe have recently undergone substantial change. The telecourse was produced before these changes occurred, but the Study Guide has been revised to reflect the current trends.

Reading Assignment

Choose the material from either textbook as your reading assignment. Your instructor might assign additional readings as well.

Living in the Environment

Chapter 28, "Politics and Environment," Sections 28-2 and 28-6
Chapter 29, "Environmental Worldviews, Ethics, and Sustainability"
Epilogue, "Principles for Understanding and Working With the Earth," pages 814-815
Guest Essay by Norman Myers, page 530

Environmental Science

Chapter 1, "Environmental Problems, Their Causes, and Sustainability," Sections 1-6 and 1-7
Chapter 2, "Economics, Politics, Ethics, and Sustainability," Sections 2-6 through 2-8
Guest Essay by Norman Myers, page 282

Unit Overview

Many scientists, policymakers, and environmentalists have concluded that the 1990s is a crucial decade, during which decisions must be made about many pressing environmental problems if the worst consequences of those problems are to be avoided. Having looked at national initiatives in the preceding unit, this unit of *Race to Save the Planet* considers responses to environmental problems at the individual and international levels. It examines the prospects for progress during what Lester Brown of the Worldwatch Institute has called "the turnaround decade."

"Now or Never" is about sustainability: how imaginative individuals can light the way toward it and how cooperation among nations can create a context for achieving it. The television program profiles individuals whose personal commitment and vision have launched important responses to environmental problems in Kenya and in Europe's Mediterranean basin. It also examines international initiatives that provide an effective framework for managing shared regional resources and responding to the complex challenge of global climate change. Examples from the Commonwealth of Independent States illustrate the rise of environmental awareness and citizen environmental activism in that nation. The program suggests that global cooperation may increase and progress toward sustainability may accelerate as East–West military tensions give way to a new concept of common security based on protection of the environment.

If you are using *Living in the Environment* as your course text, your reading assignment examines the basis for a personal environmental ethic. The chapter summarizes and contrasts various versions of two alternative worldviews—the *planetary management* worldview and the *earth wisdom* worldview. Section 29-3, called "Solutions: Living Sustainably," reviews some of the personal actions consistent with an environmentally ethical life. This material is covered in *Environmental Science* in Chapter 2, Sections 2-7 and 2-8.

This final television program in *Race to Save the Planet* confronts each of us with a single, powerful choice: the choice to become involved. Not everyone can launch a grass-roots movement like Kenya's greenbelt movement, become a Greenpeace activist, or design an initiative like the Mediterra-nean Action Plan. But everyone can begin to make small, personal decisions in their home, school, and workplace that are consistent with solutions to the problems of global change. Examples shown throughout *Race to Save the Planet* make it abun-dantly clear that the stability and resilience of the earth, and the sustainability of human society, is not a matter of "us or them." It is only a matter of "us."

Glossary of Key Terms and Concepts

The following terms and concepts will be helpful as background for viewing "Now or Never."

Agroforestry is the general name for farming practices in which trees and crops are grown in combination. Agroforestry systems benefit from the ability of trees to protect soil from erosion and to capture and cycle plant nutrients.

Glasnost (glahz-nost) is the Russian word for "openness," the policy of unrestricted public expression and discussion of controversial ideas introduced in the former Soviet Union by President Mikhail Gorbachev. Pollution and environmental

deterioration have become common topics of public debate under *glasnost.*

The **greenbelt movement** is a grass-roots tree-planting effort begun in Kenya in 1977 by Wangari Maathai and the National Council of Women of Kenya. This nationwide campaign makes tree seedlings available to individuals and community groups and raises environmental awareness through participation in reforestation.

The **Mediterranean Action Plan** is a negotiated cooperative agreement among 17 of the 18 nations that border the Mediterranean Sea (Albania is an observer) and the European Community to monitor and control pollution of the regional sea they share. Widely considered a milestone in international cooperation on the environment, the plan has so far achieved only limited success in reducing pollution levels and cleaning up the Mediterranean Sea.

The **Montreal Ozone Agreement**, signed by 24 nations in 1987 (and since then endorsed by more than 30 others), set a timetable for reducing chloro-fluorocarbon (CFC) and halon production levels by 50% by the year 2000 to control damage to the global ozone layer. Though the original schedule for reducing CFCs has already been criticized as too slow and is likely to be amended, the Montreal Ozone Agreement is considered a model of the global environmental diplomacy needed to address the more complex issue of the greenhouse effect.

Perestroika (per-uh-stróy-kuh) is the Russian word for "restructuring," the term used by ex-President Gorbachev to describe his plan to revitalize the economy of the former Soviet Union by attacking inefficiency, removing the most cumbersome aspects of centralized planning, and introducing reforms based on free-market mechanisms.

The **World Commission on Environment and Development** was established by the United Nations General Assembly in 1983 to examine international and global environmental problems and to propose strategies for sustainable development. Chaired by Norwegian Prime Minister Gro Harlem Brundtland, the independent commission held meetings and public hearings around the world and submitted a report on its inquiry, *Our Common Future,* to the General Assembly in 1987.

AFTER YOU VIEW THE TELEVISION PROGRAM

Consider What You Have Seen

"Now or Never" examines two ends of the spectrum of responses to the environmental challenges facing humanity: The program shows how individuals, acting on personal convictions, can devise imaginative and enduring responses to environmental problems, and it shows how new international forums are creating a context for regional and global environmental problem-solving. These initiatives at the individual and international levels share some common elements summarized in this unit.

- Telecourse themes
- Managing shared resources
- The role of individuals
- Devising international responses
- The elements of security

Telecourse Themes

The individuals and case studies shown in "Now or Never" illustrate a number of themes that we have encountered again and again throughout the telecourse. Take a few minutes to write down examples of each theme shown in this final television program.

Water quality

Air quality

Toxic air pollution

Point and nonpoint sources of pollution

Waste disposal

Population growth

Land degradation

Conflicting interests of industrial and developing countries

Role of information in changing awareness

Managing Shared Resources

Each of the individual and international initiatives shown in "Now or Never" involves shared resources, to which access cannot easily be restricted and damage or degradation is widely felt. The program shows cases in which resources are shared among citizens of one country (Kenya, the Commonwealth of Independent States), among citizens of many countries (the Mediterranean basin), and by the global community (the ozone layer, the potential risk of climate change). Sound management of shared or common resources invariably depends on the answers to several key questions.

Who is actually included among the affected community? In Kenya, the greenbelt movement is concerned with the soil and water resources that sustain the entire nation. But the tree-planting programs of the movement are designed especially to involve women, children, and people with disabilities, groups often excluded from conventional development activities. The Mediterranean Action Plan is an agreement that has been endorsed by 17 countries—but not all of those countries can afford to comply with the strict pollution control measures specified as desirable by the plan. The Montreal Ozone Agreement to protect the earth's ozone layer through restrictions on ozone-damaging CFCs is of concern to every nation on earth, but the success of the agreement depends on the relatively small group of nations that produce these chemicals.

What trade-offs must be made, and what conflicting interests balanced? Inevitably, management of shared resources is a process that involves trade-offs. The Neva River near Leningrad supplies both drinking water and an outfall for liquid wastes. The dam being constructed in the Neva Harbor is needed to protect Leningrad from catastrophic flooding, but it threatens to trap the city's sewage. Try to identify some of the legitimate but conflicting interests in the Mediterranean basin and in the case of global restrictions on ozone-damaging CFCs.

Where does the authority for decision making rest?   In each case of shared resources shown in the program, there is no clear responsible authority to ensure that conflicting interests are fairly balanced.   Think about who ultimately ends up making decisions about environmental quality in each case and how effectively they represent the viewpoints of the people directly affected by environmental degradation.

The Role of Individuals

"Now or Never" emphasizes the role that individuals can play in confronting environmental problems—from working to fight polluters and repair environmental damage at the local level to launching new diplomatic initiatives that foster cooperation among nations.   Think about some of the environmental leaders shown in this program and the different approaches they use.

Leading by example: Some environmental leaders change attitudes and behaviors by doing things differently themselves, using their own lives to demonstrate that change is possible.   Wangari Maathai, founder of the greenbelt movement in Kenya, has played a major and very visible role in the movement since its origins in 1977.   Her presence at tree-planting ceremonies, seedling nurseries, and communities around the country not only helps raise awareness about the need for restoration of Kenya's environment but also provides a powerful role model for girls and women in a society that does not generally encourage women's ambitions.   Denis Yatras, a Greek businessman who arranges shipments of crude oil through the Mediterranean, was moved to do something about oil pollution by his love for the sea.   Rather than abandoning his profession because it polluted the Mediterranean, he used his professional knowledge of oil trading and contacts to set up an innovative program to recover and reuse waste oil.   Scientists have often been among the leaders of change on environmental issues; like the scientists who drew attention to the deterioration of Lake Baikal in the former Soviet Union, they speak with an authority that forces reluctant politicians to take notice.

Leading by working within the system: Many other powerful figures in the environmental revolution have used existing institutions and political forums in creative new ways.   Often it takes individual vision to discern the potential to push institutions beyond their traditional limits.   Examples in "Now or Never" include Stjepan Keckes and Mostafa Tolba.   Keckes, architect of the Mediterranean Action Plan, used traditional diplomacy to bring nations together to recognize a common interest in controlling pollution of the Mediterranean. Tolba, executive director of the United Nations Environment Programme, used his UN office to raise awareness among governments about risks to the ozone layer and then host the meetings that led to the Montreal Ozone Agreement—a process that unfolded over many years.   In each case, individuals saw how existing institutions could be used to conclude unprecedented agreements on behalf of the environment.

Leading by challenging the system: Often, however, existing institutions stand in the way of solving environmental problems.   Vested interests resist environmental controls, and government and corporate bureaucracies are slow to change.   Individuals who challenge the system directly often accomplish immediate goals and also raise public awareness enough to build the political support needed to achieve far-reaching changes.

Greenpeace activists confront polluters, whalers, and nuclear-armed naval vessels; their often dramatic confrontations have a powerful symbolic value (some would say they attract too much publicity) that alerts the public to activities that typically go unnoticed.   Scientists and journalists in the former Soviet Union challenged official reports on water quality in the Neva River by conducting their own independent studies and publicizing the findings; their challenge led authorities to consider strengthening pollution controls and to reexamine the design of Leningrad's dam.   Residents of the Brateevo housing district near Moscow formed a community group that successfully pressured the Moscow City Council to reconsider plans for new industrial development near their heavily polluted district.

And Friends of the Earth–U.K., a private environmental group, successfully challenged the British government to take a stronger stand on phasing out ozone-damaging CFCs.

The examples included in "Now or Never" show that challenge can be directed at several levels, ranging from individual polluters through local authorities to national governments. Committed individuals and small groups can accomplish landmark change at each of these levels.

Leaders always face the challenge of how to get others involved, because individuals acting alone simply do not have the means to combat major environmental problems. The greenbelt movement has been especially successful in enlisting involvement of rural Kenyans. It has been effective because its tree-planting campaigns are designed to respond to local concerns—to the needs for income and firewood of the people who will plant and care for the trees. It has been successful also because it is participatory: By engagement in the planting of trees, participants learn the broader mission of the movement, raising awareness about the quality of the environment. In efforts like this, "leaders" succeed only to the extent that they put an initiative or movement squarely in the hands of the participants. This can work at many levels; the Mediterranean Action Plan could not come into force among the countries of the basin until the plan's mission had become "their" mission.

Devising International Responses

Many of the problems that must be confronted and resolved during the 1990s require international cooperative responses; even the most ambitious efforts by individual nations will not be sufficient. "Now or Never" explores several themes that bear on whether initiatives among nations have a chance of success.

Creating frameworks for cooperation: Often precedents for the sort of cooperation needed do not exist. The Mediterranean Action Plan supplied such a framework for the nations of the basin, many of whom were divided by longstanding political and cultural hostility. The plan created a context in which they could work on mutual concerns surrounding pollution, without resolving all the outstanding differences among them. The Montreal Ozone Agreement on reducing ozone-damaging chemicals is even more notable in its inclusion of a much larger group of nations. The agreement represents a true milestone in environmental diplomacy both in its demonstration that some global problems can only be dealt with at this international level and in its demonstration that the diplomatic process is capable of producing a viable agreement that nations will choose to honor and enforce.

A framework is necessary but not sufficient; it can offer no guarantee of cooperation or success. Both examples in "Now or Never" demonstrate this. The formula for phasing out CFCs, which took years to negotiate, is now believed to be too slow to avoid serious damage to the ozone layer, and the Montreal Ozone Agreement's capacity for self-correction in the face of new scientific findings has yet to be tested. And the Mediterranean Action Plan has yet to accomplish real cleanup, despite the sincere expressions of support from all parties to the agreement.

North/south issues: Global changes affect all nations, but as we have seen in earlier units, neither responsibility for the causes nor vulnerability to the consequences are equally shared. Changes in the atmosphere bring the differences among industrial and developing countries into particularly sharp focus. The meetings of the World Commission on Environment and Development provided an important forum in which perspectives of the two sets of countries could be aired. Differences concerning application of the Montreal Ozone Agreement on ozone-damaging chemicals point up some general issues likely to arise in addressing other issues that require a collective international response.

In the case of efforts to limit CFCs, developing countries have challenged the Montreal Ozone Agreement because it seems to defy the principle that those countries producing the most CFCs have to cut back the most. CFCs are integral to a range of technologies and products that developing countries associate with development: refrigeration, air conditioning, electronics, consumer products. They fear that industrial countries' profligate production of these damaging chemicals, which has caused most of the damage to the ozone layer so far, is being corrected by an agreement that requires developing countries to forfeit their hopes for modest improvements in living standards. The issue is one of equity, and to many countries the distributive aspect of the initial Montreal Ozone Agreement seems less than fair.

The best solution to this perceived inequity may not lie in redesigning the agreement so that it balances all interests (industrial and developing, past and future uses of CFCs) perfectly. The solution may lie in a more pragmatic dimension of international agreements—the question of financial assistance and technology transfer. Developing countries say that to comply with the terms of the Montreal Ozone Agreement, they need help with substitute chemicals and non-CFC technologies. If industrial countries are serious about saving the ozone layer (the developing countries say), then industrial countries have a responsibility to assist developing countries with the costliest and most time-consuming aspects of adjustment to a non-CFC world.

Think about how the global community might respond to the problem of the greenhouse effect. What sorts of apparent inequities might be inevitable in a worldwide effort to restrict the major sources of carbon dioxide emissions? What sorts of arrangements for financial and technical assistance might address developing countries' concerns about sacrificing their development aspirations because of the excesses of the industrial countries? What are some of the reasons that the greenhouse effect problem is more difficult to address than the ozone layer problem? Do you think the concerns about the CFC agreement expressed by developing countries at the London conference shown in "Now or Never" are legitimate?

The Elements of Security

"Now or Never" suggests that the next stage of the environmental revolution will be characterized by recognition that true security lies more in sustainable management of a healthy environment than in concentrations of military might. Several examples shown in the program concern progress toward this redefinition of security.

The citizens of Brateevo and Leningrad, fighting government authorities for clean air and water, challenge the idea that state-planned industrial development is worthwhile if it comes at the cost of unbreathable air and undrinkable water.

Officials from the U.S. and the former Soviet Union, gathered in Utah at a conference on climate

change, discuss the idea that reduction of tensions between the two superpowers could free up the money and brainpower needed to make tangible progress on global environmental problems.

The member countries of the Mediterranean Action Plan, by working together on the common challenge of water pollution, may begin to see unexpected opportunities for cooperation on other longstanding areas of disagreement. Actual collaboration, demanded by the nature of common environmental problems, can erode rigid ideologies and inflexible positions that have long stood in the way of peace.

How do you define "security"? As you think back on the ten television programs, what are some of the issues that nations have the best reasons to cooperate on? What are some of the concerns that you suspect will continue to divide nations and impede cooperation? This final program has suggested that successful international responses to environmental problems can be fashioned, though they are not easy to design or to enforce. But more important than the results, the process of participation, whether it involves individuals in rural communities or entire nations at the bargaining table, can and does change attitudes. Good ideas are not enough.

Take a Closer Look at the Featured Countries

Kenya, which we first visited in Unit 7, is a poor country that faces daunting obstacles to development (Figure 12-1). Chief among these is the country's population growth; the 29 million Kenyans increase by roughly 600,000 each year. The country exports tea and coffee, and its spectacular wildlife attracts tourists and revenue, but most people remain dependent on subsistence food production, livelihoods based on the health and sound management of soils and water. The gross national product per person in Kenya is about $340 per year, about one-sixtieth that of the United States. Infant and child mortality rates are high, literacy levels are low, and debilitating diseases are widespread.

Yet the country is also an important international crossroads. The capital of Nairobi

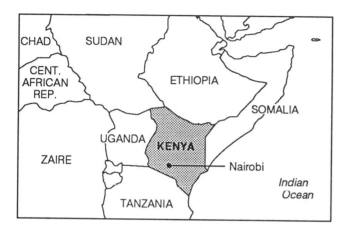

Figure 12-1   Kenya

hosts more visitors than does any other African city. The United Nations Environment Programme is headquartered there, as are a number of other important international organizations. Kenya's problems and their solutions come to typify the challenges facing Africa for many people elsewhere in the world. This means, for good or bad, that Kenya is unavoidably considered a leader in coming to grips with Africa's problems.

Most Kenyan families depend on farming, and most Kenyan farmers are women. This is true both for traditional cultural reasons and because many Kenyan men leave their families and villages in search of paying jobs in the cities. Thus leadership that emerges from the effort to confront the pressing problems in the countryside is likely to involve women. Wangari Maathai's greenbelt movement is a good example of a grass-roots movement responsive to the conditions in which most Kenyans live and to the development concerns that specifically concern women. This is perhaps the kind of environmental leadership most likely to emerge in developing countries.

In 1993 the former **Soviet Union** consisted of a loose federation of mostly industrialized countries with about 285 million citizens, expanding by about 0.6% each year (Figure 12-2). Because the country was formerly governed by a Communist regime that had traditionally been hostile to the nations of Western Europe and North America, and because the economy was managed by a centralized planning process rather than reliance on markets, economic statistics for the former Soviet Union cannot reliably be compared with those of other nations. However, the former Soviet Union's gross national product per person was estimated by various analysts at about $2,680 per year, a level about one-eighth that of the United States in 1991. A confrontational relationship with the United States since the end of World War II, known as the Cold War, was the dominant force in international politics until the recent breakup of the USSR. The two countries engaged in a nuclear arms race, devoting resources, scientific talent, and political attention to military concerns at the expense of many other priorities in both societies.

Aside from its reputation as a military superpower with a vast arsenal of nuclear weapons, the former Soviet Union had a major impact on the global environment. Heavily dependent on coal, oil, and natural gas for its energy needs, the Soviet Union was the world's second largest emitter of carbon dioxide, releasing over 1 billion tons of carbon into the atmosphere every year (roughly 18% of the world's total releases from fossil fuels). Soviet industry ranked among the least energy-efficient among industrial countries, and few industries were equipped with effective pollution controls. The world's most serious nuclear power accident occurred in a reactor at Chernobyl near the city of Kiev in April 1986, killing 31 people and putting thousands at risk to radiation-induced illness and spreading a cloud of radioactive debris across the Northern Hemisphere. The several cases shown in "Now or Never" suggest that the former Soviet Union has a long way to go on environmental protection.

The emergence of leader Mikhail Gorbachev in the mid-1980s set in motion unprecedented changes in Soviet society. By tolerating dissent and departures from the Communist party line with his policy of *glasnost*, Gorbachev encouraged an outpouring of popular feeling about pollution and environmental problems, among other formerly taboo subjects. The Chernobyl accident inspired open public opposition to new nuclear power plants. Popular participation in politics and day-to-day decision making increased. At the outset of the 1990s, the break up of the former Soviet Union and changes in its external relations could only be described as extraordinary. But the legacy of a deeply entrenched party bureaucracy, antiquated and inefficient industry, and an unresponsive system of centralized economic planning has made progress on improving living conditions in the former Soviet Union quite difficult.

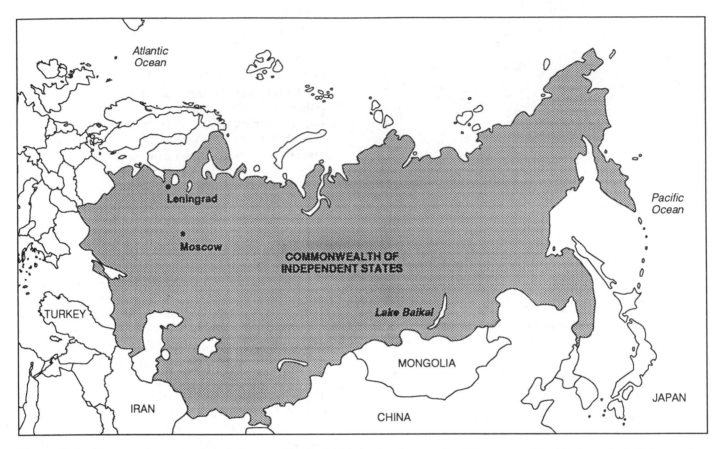

Figure 12-2   Because of recent political changes, the Soviet Union has become the Commonwealth of Independent States.

Gorbachev, and more recently Boris Yeltsin, have advanced thinking about redefining security in environmental and economic terms and backed up this thinking with unprecedented initiatives in arms control and military reductions. There was no alternative to this reordering of priorities, coupled with the gradual opening of Soviet society, if the former Soviet Union were to solve its internal environmental problems and make a contribution to confronting global problems such as climate change.

The 18 nations of the **Mediterranean basin** span an extraordinary range of cultural, economic, and demographic differences (Figure 12-3). The countries that border the Mediterranean and the island nations within it collectively contain 401 million people, growing by 1.5% or nearly 6 million people, each year. They range in size from Egypt, France, and Italy, each with more than 57 million people, to tiny nations including Cyprus and Malta with less than 1 million citizens. They range in

wealth from France, where gross national product per person is about $26,300 per year to Morocco, where GNP per person is less than $1,150 per year. Nations such as Syria and Algeria are growing at 2.8% and 2.4%, respectively, whereas Italy and Greece are essentially at zero population growth. Religious and cultural differences among nations in the Mediterranean basin are equally deep, and some countries are avowed political enemies of one another. There could hardly be a less likely group of nations on earth to come to broad, cooperative agreement on a common problem.

Yet the Mediterranean Action Plan brought this group together under a common and mutually negotiated framework to control and reduce pollution of the sea. Despite their differences, the countries of the region are united by their dependence on the sea for livelihoods and by its historical importance to their societies. None escapes the effects of marine pollution. The plan

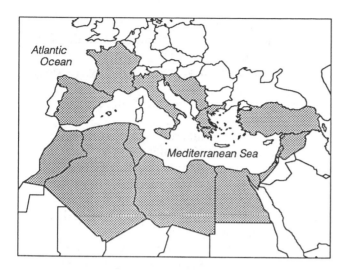

The following countries, shaded on the map, are members of the Mediterranean Action Plan:

Algeria	Italy	Spain
Cyprus	Lebanon	SyrianArab Republic
Egypt	Libya	Tunisia
France	Malta	Turkey
Greece	Monaco	Yugoslavia
Israel	Morocco	

Figure 12-3   The Mediterranean Basin

compliance still lags behind the regional has encouraged nations to collaborate in collecting scientific information about the health of the Mediterranean Sea and to curb the major sources of pollution. Not surprisingly, regulations concerning land-based pollution of the sea have been the most difficult to implement and enforce, and national agreements. The long-term goal is a regional planning process for the Mediterranean basin in which pollution control and other environmental concerns are integrated into economic decision making.

The region faces considerable changes in the years ahead, driven by changes within and among its nations. Egypt will soon overtake Italy as the most populous nation in the Mediterranean. The coastal zone as a whole will become more densely populated. Other relative demographic shifts will change the size and political importance of countries in the region. Inevitably, such changes will transform the region's environmental concerns and pose new challenges to the Mediterranean Action Plan framework. Whether the plan can

evolve along with the nations it has brought together is a challenge for the future.

Examine Your Views and Values

1. The author of your textbook draws a distinction between planetary management (or throwaway) and Earth wisdom (or sustainable-earth) worldviews. How would you describe your own worldview? Are your actions and choices consistent with that worldview?

2. Two people shown in "Now or Never" who developed innovative responses to environmental problems, Wangari Maathai of Kenya and Denis Yatras of Greece, each mentioned their personal attachment to the environment as the source of their commitment. To what in your own environment do you feel most attached? What would prompt you to act to defend or preserve it?

3. Do you think that the United States and the Commonwealth of Independent States have common interests in addressing the problem of the greenhouse effect and climate change? Would it be better for these nations to collaborate, or to compete with one another, on this issue?

TEST YOUR COMPREHENSION

Self-Test Questions
(*Answers at end of Study Guide*)

Multiple Choice

1. An important cause of soil erosion in Kenya is
    a. the fact that Kenya's young farmers do not know how to protect their land.
    b. the removal of trees for firewood and farmland.
    c. the removal of trees for charcoal used in iron smelting.
    d. naturally formed gullies caused by the country's heavy rains.
    e. More than one of the above is true. The correct answers are _____.

The correct answers are _____.

2. Women are an important focus of the Kenyan greenbelt movement's efforts because
   a. women in Kenya don't work and have time for community projects.
   b. men already understand the importance of trees.
   c. most Kenyan farmers are women and soil erosion affects them directly.
   d. women who have smaller families will be able to devote more time to the environment.
   e. More than one of the above is true.
   The correct answers are _____.

3. One of the main conclusions of the World Commission on Environment and Development was that
   a. poverty is both the cause and the effect of environmental deterioration in many developing countries.
   b. the only solution to many environmental problems will unfortunately entail an increase in poverty.
   c. rich countries are not as responsible for environmental damage as previously believed.
   d. economic growth is fundamentally at odds with environmental protection.
   e. relieving poverty will inevitably increase deforestation and soil erosion.

4. Public expression of environmental concern in the former Soviet Union was first focused on Lake Baikal because
   a. this was the most heavily polluted lake in the country.
   b. pollution of air and water had been successfully controlled in populous parts of the country.
   c. the lake was considered a national treasure because of its pristine quality.
   d. former President Mikhail Gorbachev brought it to the people's attention.
   e. foreign scientists visited the site and created an international outcry.

5. A serious new threat to Lake Baikal is
   a. new pulp and paper mills scheduled for construction in 1992.
   b. sewage from a rapidly growing city on the edge of the lake.
   c. pesticides that run off from surrounding farmlands.
   d. toxic pollutants released by routine activities of residents near the lake.
   e. toxic air pollutants released by new factories without pollution controls.

6. The residents of the Brateevo housing district near Moscow
   a. are protected from pollution by a greenbelt separating them from industrial sites.
   b. suffer from pollution but are afraid to raise their concerns with local authorities.
   c. suffer from pollution but lack the technical expertise to understand their problems.
   d. can easily move to other places when the pollution becomes intolerable.
   e. have organized and challenged local authorities to take action against polluters.

7. The Mediterranean Action Plan to control pollution in the Mediterranean basin
   a. relies on force to ensure compliance with its strict rules.
   b. was modeled on efforts to control regional pollution in other parts of the world.
   c. set an agenda for reducing releases of ozone-damaging chemicals.
   d. was one of the world's first efforts to control pollution on a regional basis.
   e. was enacted in Montreal in early 1987.

8. Scientific monitoring of pollution levels in the water is important to the Mediterranean Action Plan because
   a. traditionally there has been very little evidence that pollution was a problem.
   b. collaboration on monitoring helps build trust among the countries involved.
   c. monitoring provides baseline data used to determine appropriate responses.
   d. scientists could help interest political leaders in taking action on pollution.
   e. More than one of the above is true.
   The correct answers are _____.

9. The problem of oil pollution in the Mediterranean
   a. has now been largely solved due to the Mediterranean Action Plan.
   b. has gotten worse despite progress on other common pollution problems.
   c. is being addressed by an unusual effort to recover waste oil for reuse.
   d. is not well understood because it is so difficult to monitor effectively.
   e. More than one of the above is true.
      The correct answers are _____.

10. The most important accomplishments of the Mediterranean Action Plan so far include
    a. collection of scientific data and a framework for international cooperation.
    b. construction of sewage treatment facilities in all member nations.
    c. rigorous control of toxic pollutants in rivers that drain into the Mediterranean.
    d. ending longstanding hostilities among the member nations.
    e. More than one of the above is true.
       The correct answers are _____.

11. The main objection that developing countries had to the global protocol to cut back on production of chlorofluorocarbons was that
    a. many developing countries depend on CFC exports to pay their debts.
    b. they could not afford the more costly substitutes for CFCs.
    c. damage to the ozone layer posed very little risk to their societies.
    d. they had not been involved in negotiating the original agreement.
    e. they did not trust industrial countries such as Britain to stick to the agreement.

12. The reduction in East–West military tensions begun in the late 1980s can help nations confront environmental problems by
    a. using funds formerly spent by the military to finance sustainable development.
    b. shifting leaders' attention from military issues to broader questions of quality of life.
    c. reducing the most heavily polluting activities in industrial societies.
    d. reducing suspicions and increasing cooperation among societies.
    e. More than one of the above is true.
       The correct answers are _____.

True or False

1. About 120 nations have now signed the Montreal Ozone Agreement to cut back on CFC production. _____

2. Scientific evidence suggests that the CFC cutbacks proposed in the Montreal Ozone Agreement will not go far enough to protect the ozone layer. _____

3. Greenpeace and Friends of the Earth believe CFCs can be used safely in small quantities. _____

4. The British approach involves voluntary rather than mandatory controls on CFC producers. _____

5. The inability of nations to agree on the CFC issue makes progress on other atmospheric problems unlikely. _____

Sample Essay Questions

1. Explain the importance of participation in raising environmental awareness and improving environmental quality in Kenya's greenbelt movement.

2. Compare and contrast the Mediterranean Action Plan and the Montreal Agreement on ozone-damaging chemicals. Do you think such initiatives are significant? Why or why not?

3. How do the examples of environmental activism in Leningrad and the Brateevo housing district in Moscow illustrate the political policy known as *glasnost* (openness) in the former Soviet Union?

4. If both accept the evidence about damage to the earth's ozone layer, why do industrial and developing countries have different positions with regard to the speed and method of phasing out CFCs?

5. Propose an agenda of environmental priorities the United States could address with resources made available by reductions in military spending. What should United States policy toward the Commonwealth of Independent States be as it pursues these new environmental goals?

GET INVOLVED

References

Athansiou, Tony. *Divided Planet: The Ecology of Rich and Poor*. Boston: Little, Brown, 1996.

Bossel, Harmat. *Paths to a Sustainable Future*. New York: Cambridge University Press, 1998.

Brundtland, Gro Harlem. "Essay: How to Secure Our Common Future." *Scientific American*, Sept. 1989.

MacDonald, Mary. *Agendas for Sustainability: Environment and Development in the 21st Century*. New York: Routledge, 1998.

Markley, Oliver, and Walter R. McCuan, eds. *21st Century Earth: Opposing Viewpoints*. Boston: Greenhaven, 1996.

Myers, Norman. *Ultimate Security: The Environmental Basis of Political Stability*. New York: Norton, 1996.

Switzen, Jacqueline, and Gary Byner. *Environmental Politics: Domestic and Global Dimensions*. New York: St. Martin's, 1998.

World Commission on Environment and Development. *Our Common Future*. New York: Oxford University Press, 1987.

The following chapter in *Taking Sides: Clashing Views on Controversial Environmental Issues* takes a close look at the Montreal Ozone Agreement.

- Issue 18. "Does the Montreal Ozone Agreement Signal a New Era of International Environmental Statesmanship?" (pp. 308–325)

Also see chapters in various recent editions of Worldwatch Institute's annual *State of the World* report that deal with redefining security in environmental terms.

- Michael Renner, "Ending Violent Conflict," *State of the World 1999*.

- Michael Renner, "Budgeting for Disarmament," *State of the World 1995*.

- Hilary F. French, "Forging a New Global Partnership," *State of the World 1995*.

- Sandra Postel and Christopher Flavin, "Reshaping the Global Economy," *State of the World 1991*.

- Michael Renner, "Converting to a Peaceful Economy," *State of the World 1990*.

- Lester R. Brown et al. "Outlining A Global Action Plan," *State of the World 1989*.

- Michael Renner, "Enhancing Global Security," *State of the World 1989*.

- Lester R. Brown and Edward C. Wolf, "Reclaiming the Future," *State of the World 1988*.

Organizations

The following organizations are only a few of those involved with the kinds of initiatives shown in "Now or Never." Greenpeace, the largest member-supported environmental group in the world, describes its mission as "to personally bear witness to atrocities against life, [and] to take direct nonviolent action to prevent them." The Centre for Our Common Future was created to publicize and to follow up on the work of the World Commission on Environment and Development.

Centre for Our Common Future
Palais Wilson
52 Rue de Paquis
CH-1201 Geneva, Switzerland

Greenpeace USA
1436 U Street, NW
Washington, DC 20009
Tel: (202) 462-1177; web: www.greenpeaceusa.org

Beyond the Year 2000

These children of Niger's Majjia Valley will ultimately benefit from today's wise resource management. Building windbreaks will help to protect this arid valley from erosion and ensure healthier crops for the future.

TIME TO REFLECT

Learning Objectives

After completing the assigned readings and reviewing the material in earlier units, you should be able to

- explain why some global changes in coming decades are nearly certain while others are speculative and what this difference in levels of certainty has to do with how we respond to environmental problems.

- explain what is meant by "double-benefit" responses to environmental problems and list some of the environmental problems for which such responses are appropriate.

- describe some of the steps that governments and individuals can take to address major environmental challenges during the 1990s.

- outline how you plan to stay involved with issues presented in *Race to Save the Planet* after the end of this course.

Reading Assignment

Choose the material from either textbook as your reading assignment. Your instructor might assign additional readings as well.

Living in the Environment

Epilogue, "Principles for Understanding and Working With the Earth," pages 814-815

Environmental Science

Chapter 2, "Economics, Politics, Ethics, and Sustainability," Sections 2-7 and 2-8.

"Solutions: Principles of Sustainability: Learning from Nature," page 176

Unit Overview

This concluding print-only unit of *Race to Save the Planet* gives us a chance to deepen our understanding of sustainability by reviewing common principles woven throughout the preceding twelve units and considering how the world beyond the year 2000 may be shaped by our responses to environmental problems in the 1990s.

The epilogue in your text provides a brief summary of the assumptions on which the book is based, a distillation of the sustainable-earth worldview. This perspective emphasizes complexity and interconnectedness, the need for human activities to take heed of the energy flows and materials cycles of the biosphere, and the importance of individual conviction and action. Your instructor may assign additional readings that discuss how present choices about technology, energy use, and resource management will influence the shape of the future.

The remainder of this final unit considers how the environmental revolution presented in the television programs is likely to unfold in the years ahead. We will review trends and developments that are certain to take place, examine changes that are less certain but likely, and attempt to gauge the dimensions of unknown and unforeseeable events that will shape our future. We will evaluate some important steps that individuals and governments can take during the 1990s, emphasizing themes from the ten television programs. And we will briefly review the countries visited in the ten programs, to see whether the perspective we have developed in the course is a global one. Finally, the Study Guide urges you to stay informed through further reading—and stay involved by deepening your understanding of how the issues presented in *Race to Save the Planet* affect and are affected by your own life.

William D. Ruckelshaus, former administrator of the U.S. Environmental Protection Agency and a member of the World Commission on Environment and Development, has written that our ability to complete the environmental revolution will depend on the emergence of a "sustainability consciousness" in societies around the world, a way of thinking that is consistent with the many adjustments individuals and societies must make. The 1990s is the decade in which this new way of thinking will take shape. It can begin—it must begin—with you.

THE ENVIRONMENTAL REVOLUTION UNFOLDS

The Shape of the Future

Throughout *Race to Save the Planet*, we have been concerned with the dynamics of change: the nature of exponential growth, how rates of change have varied throughout human history, how technologies and population trends affect change. The shape of the human future depends on changes that are underway today. By learning to relate present trends to the future, we can begin to see how actions and decisions today can influence the world a generation hence.

Despite all the computers at our command today, we can be no more *certain* about the future than were the ancients who tried to divine their destiny by reading the entrails of birds. Unlike the ancients, however, we can distinguish changes that *are* certain over some specified time span from those that are merely likely and those that cannot be foreseen. This can help clarify actions that must be taken to avoid undesired consequences and guard against unrealistic expectations of quick solutions. Looking beyond the year 2000, what can we discern about the condition of the earth and of *Homo sapiens*?

Certain changes include the dimensions of phenomena that currently exhibit exponential growth, chief among them the growth of the human population. Demographers make projections about future population size based on present trends and differing assumptions about how fast fertility levels will decline. The projections vary, but all agree on one thing: The human population will continue to grow larger for several decades. Our numbers will grow to six, seven, then about eight billion by the year 2020; this much is inherent in the age structure and fertility levels of the present population.

Commitments to family planning can alter the ultimate size of the population, bring much closer the day that growth will level off, and improve the well-being of women and children now living, but they cannot stop our growth at anywhere near today's 5.7 billion.

Changes in the chemistry of the atmosphere are equally certain. As we learned in Unit 4, the long lifetimes of the chlorofluorocarbons (CFCs) mean that those we have already released to the skies will continue to damage the ozone layer for a century or more. We can limit the *extent* of damage by halting production and use of CFCs, as revisions to the Montreal Ozone Agreement aim to do, but the reality of the coming century is life under a somewhat reduced ozone layer. Likewise, the accumulation of the greenhouse gases carbon dioxide, methane, and various industrial gases, linked to population growth, affluence, and development, can be slowed but not (at least in the next several decades) halted or reversed. Reliance on fossil fuels and carbon-releasing activities is too widespread in the present world. We no longer have a choice about whether or not to alter the composition of the atmosphere and risk changes in the earth's climate; our choice is limited to the extent of the change and the rate at which it will unfold.

Other changes driven by an exponential dynamic confront us with the same sort of certainty. As long as forests and other natural ecosystems are cleared for farmland and other uses, particularly in tropical countries where population growth and the need for land are greatest, large numbers of plant and animal species will become extinct. We cannot say which ones or how many, so incomplete is our knowledge of the earth's biological diversity, but we can say that saving *all* species now living on earth is not among the possibilities open to us. Another such problem is the accumulation of debt by developing countries; as long as they must pay current obligations to their creditors by borrowing more money, their total debt burden will grow, the period needed for repayment will lengthen, and the likelihood of repayment will diminish. There is no simple way to wipe the slate clean, and the reality of enormous debt will burden countries' development prospects, and the world economy, far into the future.

Because these sorts of changes are certain does not mean the situation is hopeless, only that we must approach our environmental problems with a clear sense that we do live in an age of global change, a time of transition for the planet. Our responses to phenomena such as human population growth and atmospheric change can influence the *rate* of change and its ultimate *magnitude*—our two most important ways of adapting to a changing world.

Most elements of the world beyond the year 2000, of course, cannot be forecast with such certainty. But many **likely changes** can be anticipated—ways in which we can be relatively confident that the world in the near future will differ from our present world. Many of the trends and developments we have covered in *Race to Save the Planet* can have lasting impact on humanity's ability to respond to environmental problems and to enact sustainable forms of development. Many involve innovative technologies or institutional arrangements that, although represented by only a few examples today, are likely to proliferate in the years ahead.

Some of the changes likely to have important lasting consequences are unfolding on the political scene. The reduction of tensions between East and West, begun in the late 1980s and accelerated by the unprecedented success of popular democratic movements in Eastern Europe and parts of the former Soviet Union in 1989, has changed suspicion and hostility, backed up by troops and nuclear weapons, into cautious cooperation. The post–Cold War world offers national leaders an opportunity to shift attention and resources from military concerns to broader concepts of mutual and global security— a political climate in which environmental concerns are almost certain to rise in importance.

This shift in attitude is reinforced by increasing integration and interdependence in the world economy, which will be marked during the 1990s by events such as the consolidation of national economies in western Europe into a single regional unit in 1992 and the growing importance of regional coordination in Asia, parts of Africa, and elsewhere in the world. The more nations collaborate in economic affairs, the more they may find common ground on environmental problems.

Political and economic developments like these should brighten prospects for the sort of international environmental diplomacy represented by the Montreal Ozone Agreement on CFCs and the Mediterranean Action Plan. It is probably safe to expect to see more regional and global agreements on environmental matters, including treaties and conventions on such challenges as climate change and the loss of the earth's biological diversity. Negotiating such agreements is a slow and uncertain process, but unambiguous evidence of physical changes (including damage to the ozone layer) may speed it up. Countries unwilling to make sacrifices on their own (in energy use, for example) may agree to do so within a common framework for sharing the burdens.

Other likely changes include the spread of technologies and responses to environmental damage that have now demonstrated their potential in at least a few places. A shift toward more reliance on *input* methods to control pollution seems likely, as the economic case for pollution prevention is made clearer. A trend toward waste reduction by capturing and reusing materials in industry, like the trend toward low-input approaches in farming, is driven by a powerful economic rationale. Firms such as the Aeroscientific Corporation (Unit 10) and farms like Fred Kirschenmann's in North Dakota (Unit 9) could become the rule rather than the exception during the 1990s. Though the collective impact of such changes may be impossible to predict, they represent solid progress toward sustainability.

Much about the future remains **unknown** and **unforeseeable**. Events and developments that we have no way to anticipate could shape both global changes themselves and the human response. New technologies may fundamentally change the way humanity uses energy or produces food. Natural catastrophes and unexpected technological failures can have wide-ranging environmental and human consequences. Political events can change the course of history—and with it the willingness and ability of societies to solve national problems and respond to global changes. In addition, many of the consequences of well-recognized trends cannot be foreseen. Though the change in the composition of the atmosphere is beyond dispute, the timing and character of the climate change that will result remain great mysteries. The boundaries of uncertainty are wide.

Among the greatest unknowns is the degree to which individuals and nations can modify deeply held values that are at odds with progress on global change. Much of the environmental revolution will depend on individuals choosing to see the world differently and live their lives differently—to live more in accordance with a sustainable-earth worldview.   The values on which such a perspective rests can be explained, illustrated, and taught, but the decision to honor those values is one that must be made separately in millions of households and workplaces around the world.

Choices for the 1990s

Within this mix of certain, likely, and unforeseen changes, the future habitability of the earth will depend largely on decisions and commitments made during the 1990s.   The environmental revolution will be a composite of many separate but related revolutions in energy use, sustainable-farming practices, waste reduction and management, and so on, and each of these will depend on choices made by both governments and individuals.   One criterion that has been suggested to help in deciding what actions to take during the 1990s is to do those things that meet a double-benefit standard: actions or investments that will prevent or delay the most serious future consequences *and* yield direct benefits in the present.

Some of the key steps that can be taken during the 1990s by governments and individuals are summarized below.

Population

Governments can set national goals for achieving stable populations, support national family-planning programs, and provide adequate health-care services, particularly for women and children.

Individuals can learn family-planning alternatives; decide what family size is appropriate for their level of affluence and what the impact of an additional child might be; and support groups that provide family-planning education and services, both domestically and internationally.

Atmospheric Change

Governments can decide to accelerate national phase-out of chlorofluorocarbons (CFCs); set a timetable for planned reductions in fossil-fuel use and a strategy for accomplishing them; support international efforts to set limits on the release of

greenhouse gases; and fund research on climate change and its implications for their societies.

Individuals can reduce personal reliance on fossil fuels by eliminating energy waste at home and in the workplace; and they can support organizations active in public education on climate change.

Pollution of Air and Water, Toxic Waste

Governments can enforce compliance with output controls on pollution sources; support adequate funding for cleanup of toxic waste dumps; encourage industries, businesses, and households to practice input controls and make input control the centerpiece of new pollution control policies; and publicize successful pollution reduction efforts by government and private enterprises.

Individuals can learn about local pollution issues and get involved in local politics to influence decision making; support groups advocating wider reliance on input controls; and inventory household toxic materials and outline steps for disposing of them safely.

Ecosystems and Biological Diversity

Governments can expand national systems of protected natural areas and involve local communities in managing them; and they can promote alternatives to destructive uses of forests and wildlands that eliminate species and damage ecosystems.

Individuals can learn about local ecosystems and the plant and animal species characteristic of them; and they can support organizations that study and protect threatened ecosystems and endangered species.

Energy

Governments can seek to limit use of fossil fuels by taxing fuels according to their contribution to the greenhouse effect; support public transport in urban areas and between cities; make energy efficiency the centerpiece of national energy policies; and support research and development of energy-efficient technologies and renewable energy sources.

Individuals can reduce energy use by conserving energy at home; buying energy-efficient models to replace old appliances; reducing unnecessary travel by automobile; and using public transit or bicycles when possible.

Farming and Food Production

Governments can support research and extension to farmers on low-input sustainable methods of farming; review national farm policies and phase out subsidies that hide the true costs of water, energy, and agrochemicals; publicize the results of surveys of land and water contamination by agrochemicals; and examine the environmental impact of policies that promote surplus production and exports.

Individuals can research sources of food in local stores and stay abreast of food-safety issues; support local growers by buying at farmers' markets and stores that carry local produce; and produce food at home in a backyard or community garden.

Solid Waste

Governments can review the costs of landfill disposal of garbage and subsidize recycling efforts that represent a cost saving; support the development of local and regional markets for recycled materials; and encourage businesses and industries to practice source reduction of solid waste.

Individuals can learn where their garbage is deposited; inventory household waste to see how its volume can easily be reduced; and participate in voluntary or community recycling programs.

Taking the steps summarized here involves a reappraisal of development paths in both industrial and developing societies. Throughout *Race to Save the Planet* we have emphasized that the environmental revolution demands far-reaching adjustments in all societies. Industrial countries need to address both the obvious and the subtle costs of overdevelopment and consumption overpopulation and to ask how their societies can be sustained with less risk to the stability of the global environment. Developing countries need to find ways to address poverty and people overpopulation and to meet the needs and aspirations of their people without repeating the mistakes and reinforcing the damage caused by industrial societies. Sustainable

development is a concept and a standard meant to apply to societies at all levels.

The concept of sustainability that underlies this new approach to development attempts to put the present and the future on equal terms. According to the definition adopted by the World Commission on Environment and Development in *Our Common Future*, "Sustainable development seeks to meet the needs and aspirations of the present without compromising the ability to meet those of the future." Living as if the future mattered, rather than as if it might never arrive, is the heart of the sustainable-earth worldview. The most important actions and decisions of the 1990s will be those that reflect this view.

WHERE WE HAVE BEEN

A Geographic Summary

In the course of the ten programs of the *Race to Save the Planet* series, we visited 29 different countries (17 of which have been profiled in this Study Guide) and the continent of Antarctica, and focused on 4 distinct regions (the Los Angeles basin, the Rhine basin, the Mediterranean basin, and the Sahel region of Sub-Saharan Africa). How well has this set of places helped us build a global perspective?

Of the 29 countries, 21 were discussed in the Study Guide; these countries contain about 43% of the world's population and represent most regions on earth. However, the coverage of different regions varies considerably. Athough the five African countries featured illustrate the geographic range of that continent, these countries contain just 8% of Africa's people. The two other major developing regions are better represented, with featured countries containing roughly 40% of the population of both Latin America and Asia. Europe is more thoroughly depicted by the programs; featured countries (including the nations of the Rhine basin) account for nearly two-thirds of the population of northern and western Europe.

The world's most affluent societies (the United States, western Europe, and Japan) and some of its poorer societies are shown. But the conditions under which the vast majority of the world's people

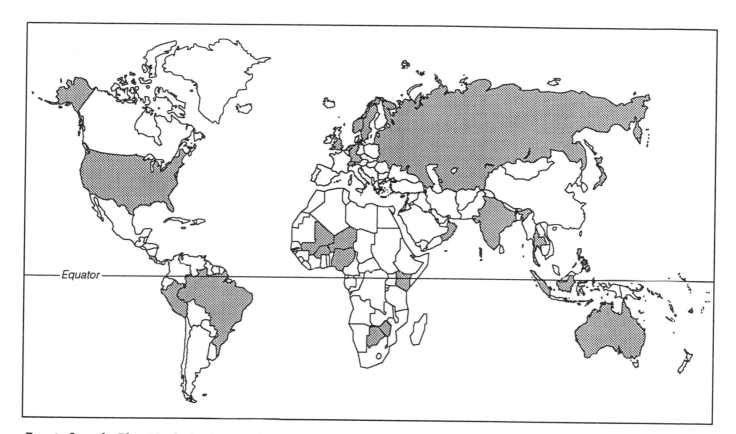

Race to Save the Planet includes footage from the following 29 countries:

Australia*	Denmark*	Jordan	Norway	Sweden*
Botswana	England	Kenya*	Oman*	Switzerland
Brazil*	India*	Mali	Peru*	Thailand*
Burkina Faso	Indonesia*	The Netherlands*	Philippines	United States*
Costa Rica*	Israel	Niger	Former Union of Soviet	Former West Germany
Djibouti*	Japan*	Nigeria	Socialist Republics*	Zimbabwe*

* Featured in Study Guide country profiles

Figure 13-1

live are not treated consistently throughout the series; as has been true throughout human history, the well-to-do remain more visible than the poor.

Another way to evaluate this roster of countries is to consider their importance to global environmental challenges. Eight of the world's top ten carbon dioxide emitters (and contributors to the greenhouse effect) are among the countries featured; these countries will play central roles in responding to the risk of climate change. Four of the ten countries containing the greatest diversity of plant and animal species are included as well. By at least these two indicators of importance on major environmental issues, the television programs

provide a global perspective.

What regions and countries are left out? Perhaps the most important omission is the Middle East, despite the fact that some of the countries in that region sit atop the world's largest remaining reserves of crude oil. China is also excluded, although the world's most populous nation and third largest emitter of carbon dioxide obviously will play a prominent role in the global future. Finally, the filmmakers have included no Eastern European countries. Although environmental problems in the former Soviet Union and the rise of environmental activism there to some extent parallel those issues in Eastern Europe, political environmental activism there to some extent parallel those issues in Eastern Europe, political change in that region and its reliance on heavy industry mean that events in Eastern Europe will be an increasingly important part of the global environment picture.

No series of television programs could hope to portray the full diversity of the more than 160 nations on earth or to present a balanced portrait of all regions. However, the countries featured in *Race to Save the Planet* provide a thorough introduction to the world's economic and cultural spectrum and the range of responses to environmental problems arising in societies around the world. Without feeling that you have seen it all, your appreciation for the world's variety should be deepened by the geographic scope of the programs.

Gauging Your Comprehension

As we reach the end of the course, you may want to know how you can evaluate your mastery of the material we have covered before you take a final exam. A good way to begin is by reviewing the learning objectives of each of the preceding 12 units. You may even want to photocopy the pages of this Study Guide listing those objectives and highlight or circle the ones that you feel least comfortable answering. Use these priorities to decide how to focus your review of each unit. Also, complete any reading assignments you may have skipped or postponed earlier in the course.

Although understanding the principles, concepts, and examples from each unit will be important on a final exam, the most important test of your comprehension is how well you can relate the topics covered in the ten television programs to your daily activities. By this point, you should be able to discuss how any of the topics covered in *Race to Save the Planet* touch your life and why your choices matter. You should be able to explain what the problems of ozone depletion and solid waste disposal, for instance, mean to you and how you can address the causes. If you have integrated at least some of these concerns into your personal life, you have fulfilled the hopes of the creators of this telecourse.

WHERE ARE WE GOING?

Stay Informed

Each unit of this Study Guide has listed books and articles selected to deepen your understanding. This unit concludes with a short list of references that deal with the future: what it might look like, how to think critically about it, and what to do about it. You will find a much more comprehensive list of references on all the topics covered in *Race to Save the Planet* at the back of your textbook. To stay abreast of global environmental issues, you may also want to subscribe to one or more of the periodicals listed below.

References

Ausbel, Kenny. *Restoring the Earth: Visionary Solutions from Bioneers.* Tiburon, CA: H. J. Kramer, 1997.

Hardin, Garrett. *Filters Against Folly.* New York: Viking Penguin, 1985.

Learner, Steve. *Eco-pioneers: Practical Visionaries Solving Environmental Problems.* Cambridge, MA: MIT Press.

Leopold, Aldo. *A Sand County Almanac.* New York: Oxford University Press, 1949.

Millbrath, Lester W. *Learning to Think Environmentally While There Is Still Time.* Albany: State University of New York Press, 1995.

Quinn, Daniel. *Ishmael.* New York: Bantam/Turner, 1992.

Ruckelshaus, William D. "Toward a Sustainable World." *Scientific American*, Sept. 1989.

Seymour, John, and Herbert Girardet. *Blueprint for a Green Planet: Your Practical Guide to Restoring the World's Environment.* Englewood Cliffs, NJ: Prentice-Hall, 1987.

Suzuki, David. *The Sacred Balance: Rediscovering Our Place in Nature.* New York: Prometheus, 1998.

World Commission on Environment and Development. *Our Common Future.* New York: Oxford University Press, 1987.

The following chapter in *Taking Sides: Clashing Views on Controversial Environmental Issues* contrasts cornucopian and neo-Malthusian views about the future.

. Issue 19. "Are Abundant Resources and an Improved Environment Likely Future Prospects for the World's People?" (pp. 326–343)

You may find the following chapters from the Worldwatch Institute's annual *State of the World* reports helpful.

. Lester R. Brown "Challenges of the New Century." *State of the World 2000*

. David Malin Roodman, "Building a Sustainable Society." *State of the World 1999*

. Lester R. Brown and Jennifer Mitchell, "Building a New Economy." *State of the World 1998*

. Lester R. Brown, Christopher Flavin, and Sandra Postel, "Picturing a Sustainable Society." *State of the World 1990*

Periodicals

World Watch
Worldwatch Institute
1776 Massachusetts Avenue NW
Washington, DC 20036

An excellent bimonthly source of readable, in-depth articles on many of the topics introduced in *Race to Save the Planet*, written from a global perspective.

Environment
Heldref Publications
1319 Eighteenth St. NW
Washington, DC 20036-1802

A scholarly monthly journal on current environmental problems with an emphasis on public policy, covering both domestic and international issues.

Orion Nature Quarterly
136 East 64th Street
New York, NY 10021

A quarterly journal devoted to exploring all aspects of the relationship between humanity and the natural world; strong emphasis on environmental values and the ethic of stewardship.

This short list only begins to suggest the variety of high-quality periodicals on environmental topics. Many are published by major environmental groups. Check your community or university library for some of the following: *Amicus Journal, Audubon, E, Earth Island Journal, Greenpeace Magazine, International Wildlife, National Wildlife, Not Man Apart, Sierra,* and *Wilderness.*

Science magazines such as *Discover, Science News, Natural History,* and *Scientific American* offer comprehensive regular coverage of environmental topics. In addition, major newsweeklies such as *Time, Newsweek, U.S. News & World Report,* and even *Sports Illustrated* often have first-rate reporting of current environmental news.

Stay Involved

In Unit 1, you were asked to write a profile of your lifestyle before viewing any of the television programs. Take time to look back at what you wrote at the start of the course, and note the issues or behaviors that you see differently as a result of what you have learned. Identify changes in your present way of living that could do the most to help address environmental problems. For some suggestions, see Appendix 5 in *Living in the Environment* and Appendix 6 in *Environmental Science.*

This course is not over when you complete the last unit and take a final exam. After taking a second look at your lifestyle profile, list some of the steps you might take over the course of the next week/month/year to continue your involvement and increase your understanding of the issues we have covered in *Race to Save the Planet.* For example, you might choose to:

Within the next week:

- subscribe to an environmental magazine and join a local or national group active in one of the areas that most interests you

Within the next month:

- gather information about stores in your area that sell energy-efficient appliances

- identify local sources of chemical-free produce

- find out whether your local utility sponsors energy conservation programs in which households can participate

Within the next year:

- consider replacing your current refrigerator or water heater with an energy-efficient model

- replace burned-out light bulbs with compact fluorescent bulbs

- buy a bicycle to use as a daily or occasional alternative transport to your work

Each of us can make a crucial difference, and every contribution counts. However you choose to stay involved, you will be part of the race, helping to secure the future of the earth.

Answers to Self-Test Questions

Unit 1

Multiple Choice

1. d. exponential growth.

2. d. the production of automobiles by a single factory

3. a. leads to environmental degradation and makes the resource nonrenewable.

4. c. input and output methods.

5. b. concentrated forms capable of performing work.

6. c. says that energy input always equals energy output.

7. b. means that some quantity of waste will always be produced.

8. e. is a demonstration of the second law of thermodynamics.

9. e. input approaches would be used to reduce materials waste and prevent pollution.

True or False

1. Ninety-five percent of the energy used in a typical incandescent bulb is lost as heat.   (true)

2. The most efficient way to heat bath water is to use electricity generated in nuclear power plants. (false)

3. Energy efficiency saves energy at more than the cost of generating it. (false)

4. Only 16% of the commercial energy used in the United States actually performs useful work. (true)

5. Superinsulating a home is the most efficient way to provide heat in a cold climate. (true)

6. Net useful energy is the same as the potential energy from a fuel source. (false)

Unit 2

Multiple Choice

1. d. less than 1%; the rest is reflected or absorbed as heat.

2. a. differences in average temperature and average rainfall.

3. e. (a, c)

    a. converts carbon dioxide and water in the presence of sunlight into glucose and oxygen.

    c. is carried out by organisms known as producers or autotrophs.

4. b. limiting factor principle.

5. b. more trophic levels means a greater loss of high-quality energy from the system.

6. c. the amount by which the energy captured in photosynthesis exceeds the energy used in respiration by plants.

7. b. burning carbon-containing fossil fuels and destroying forests.

8. b. the changes are likely to affect the earth's temperature.

9. d. desert.

10. d. most of its nutrients are in the vegetation rather than the soil.

11. b. tallgrass prairie.

12. e. natural process that human activities often accelerate.

True or False

1. Major ecosystem types found on land are known as "ecotones." (false)

2. A biological community includes the plants, animals, and decomposers found in a particular place, but not the nonliving elements. (true)

3. A species includes all populations of a particular organism that can interbreed and produce fertile offspring.   (true)

4. Organisms called consumers or heterotrophs can manufacture their own food.   (false)

5. Decomposers assist the recycling of chemical elements within ecosystems.   (true)

6. Most biogeochemical cycles are too vast to be affected by human activities.   (false)

7. The carbon cycle involves only the atmosphere and living organisms.   (false)

8. The carbon cycle involves a linkage between photosynthesis and aerobic respiration.   (true)

9. The nitrogen cycle is important to living organisms because nitrogen is found in proteins and DNA molecules.   (true)

10. Nitrogen fixation is carried out by most plants but only some microorganisms.   (false)

11. Water is transferred from the oceans to land in the hydrologic cycle by the power of the sun.   (true)

12. The oceans account for a larger share of the earth's total net primary productivity than does any other ecosystem type.   (true)

13. Farmland has a higher net primary productivity than does any terrestrial ecosystem type except the rain forest.   (false)

14. Because tropical forests are very productive, they make the best farmland when cleared.   (false)

15. Harvests can diminish the productivity of any ecosystem by removing nutrients.   (true)

16. Forests may contain more animal species than do grasslands because their complex structure supplies more distinct niches.   (true)

Unit 3

Multiple Choice

1. b. the mobility required for survival keeps fertility rates low.

2. c. fertility levels rose in settled communities and populations began to increase.

3. a. settled hunting and gathering societies could overuse their environment.

4. e. they gradually exhausted supplies of game and wild plants.

5. c. about as fast as populations we would consider "stable" today.

6. d. draining underground coal mines using steam-powered pumps.

7. b. marriage rates and fertility rates responded to living standards.

8. d. it provided the first global forum for dialogue on environmental problems.

True or False

1. England's shift to reliance on coal was prompted by the exhaustion of wood supplies.   (true)

2. When the Industrial Revolution began in England, the country had not yet sustained any environmental damage.   (false)

3. Pollution by the new industries in England was welcomed by many people as a sign of prosperity.   (true)

4. The first evidence of pollution in England was declining game populations.   (false)

5. Workers joined the new industries because farming was dying out.   (false)

6. Changes in agriculture were as important as changes in manufacturing methods in the Industrial Revolution. (true)

7. The two key developments that encouraged the environmental awakening of the 1960s were fallout of strontium-90 from bomb testing and publication of *Silent Spring*. (true)

8. Although Rachel Carson's book was influential, her claims about pesticides were proved to be exaggerated. (false)

9. Once people recognized the dangers of DDT, the pesticide was quickly banned. (false)

10. The Yannacones' lawsuit to ban spraying of DDT on Long Island was the first significant use of the courts for environmental protection. (true)

11. Earth Day 1970, though a popular success, was ignored by most politicians. (false)

12. The environmental movement in the United States was based on progress against pollution in Europe. (false)

Unit 4

Multiple Choice

1. e. (b, c)

   b. increased risk of skin cancer.

   c. higher than usual levels of ultraviolet light from the sun.

2. c. ultraviolet light from the sun.

3. c. increased by 30%.

4. b. is likely to be reached by the year 2050.

5. c. the computer models used to predict future climate changes can also account for past climate changes.

6. e. (c, d)

   c. is likely because water expands as it warms.

   d. may be caused by melting ice at the North and South poles.

7. c. the costs of protection from rising seas would be beyond the reach of some nations.

8. e. (c, d)

   c. towns, farms, and roads would prevent migrating animals from relocating.

   d. the area might be the first park to succumb to the greenhouse effect.

9. a. is thought to have a slight cooling effect.

10. c. "scrubbers" to catch carbon dioxide in smokestacks.

True or False

1. Flooded rice paddies are an important source of carbon dioxide. (false)

2. The major source of carbon dioxide is burning fossil fuels. (true)

3. Deforestation causes carbon dioxide to be absorbed from the air. (false)

4. Methane counteracts the effect of carbon dioxide in the atmosphere. (false)

5. CFCs contribute to both global warming and ozone loss. (true)

6. In 1988, the U.S. grain harvest was reduced by nearly half. (true)

7. Pollination of corn plants is more efficient at high temperatures. (false)

8. The greenhouse effect would compensate for diminished flow of the Colorado River by causing more rain in the U.S. Southwest. (false)

9. The system to allocate water to users from the Colorado River is not designed for periods of lower flow. (true)

10. Few major rivers in the world are used to capacity. (false)

Unit 5

Multiple Choice

1. b. nitrogen oxides and hydrocarbons from cars and industries interacting with sunlight.

2. c. it causes the most obvious breathing problems and other health-damaging effects.

3. c. the South Coast Air Quality Management District plan.

4. c. there was a measurable increase of pathogenic occurrences.

5. d. a new catalytic converter for automobiles.

6. c. nutrients from farmland and other nonpoint sources throughout the basin.

7. e. (c, d)

   c. produced less pollution than the river receives each day from usual sources.

   d. caused an enormous increase in public awareness of pollution of the Rhine.

8. a. fish may reveal the effects of previously unsuspected pollutants.

9. c. are increasing faster than they can be monitored by scientists.

True or False

1. Acid deposition is not a problem in Los Angeles, because there are few industrial sources of sulfur oxides. (false)

2. Although the health-damaging effect of air pollutants is well known, damage to trees and vegetation in the Los Angeles basin has been observed only recently. (false)

3. People in Los Angeles become physically more tolerant to pollution exposure. (false)

4. The major source of smog-causing pollutants in Los Angeles is automobiles. (true)

5. The South Coast Air Quality Management District plan is based on approaches that have proved successful in other U.S. cities. (false)

Unit 6

Multiple Choice

1. e. (b, d)

   b. very similar to the model of development followed by countries in Europe and North America.

   d. in which industrialization plays a major role.

2. e. (a, c)

   a. must begin with the impoverished families in India's 700,000 villages.

   c. could be based on small-scale industries that supplied local people with essentials.

3. c. have been moved to resettlement villages because their land was needed for the complex.

4. e. (a, c)

   a. the Singrauli complex proves that the trickle-down theory of development doesn't work in India.

   c. the Singrauli complex could have been more beneficial if it had been managed by environmentalists.

5. c. is steep and prone to soil erosion.

6. e. (b, c)

   b. women's activities are directly dependent on the health of forests and the environment.

   c. women are leaders of Chipko as well as followers.

7. c. may be the most populous nation on earth by the turn of the century.

8. e. (a, b)

   a. children bring income and security to their parents.

   b. children can help with farm work.

9. e. (a, b, c, d)

   a. the largest slum in Latin America is in Rio de Janeiro.

   b. the rain forests of the Amazon basin are being destroyed.

   c. children in Brazil still die of preventable diseases such as measles.

   d. Brazil has the largest external debt in Latin America.

10. b. is an industrial development scheme that includes iron mines, sawmills, ranches, and plantations.

11. b. could lead to complete deforestation by 2010.

12. e. (a, b)

   a. the rubber tappers' livelihood was threatened by ranchers, farmers, and miners invading their forest areas.

   b. powerful landowners used violence to intimidate the rubber tappers.

Fill In

1. The thermal power complex at Singrauli poses a number of environmental problems. Although the coal-fired plants are equipped with (scrubbers), these control only (particulates), not (sulfur dioxide) or (nitrogen oxides). Local water supplies and adjacent farmland have also been contaminated by (ash) from the plant.

2. The grass-roots movement based on protecting forests that arose in the Himalayas is called (Chipko). Its leaders were inspired by Gandhi's example of (nonviolence). The forests that the movement's followers sought to protect supplied them with (fuel or firewood), (fodder), and (building materials).

3. The people who gather latex in the western Amazon region are known as (rubber tappers). Unlike other forms of development in the Amazon, their livelihood does not cause (deforestation) and is considered by many to be a model of (sustainable development). Their work is threatened by (farmers), (ranchers), and (miners) who come to the Amazon in search of land.

Matching

1. Prime Minister Nehru   b, d

2. Mahatma Gandhi   i

3. Singrauli, India   b, d, e, f, h

4. Chandi Prasad Bhatt   a, c, i, j, k

5. Chico Mendes   c, k, l, o

6. World Bank   b, d, n

7. Amazon basin   g, l, m, o

8. Greater Carajas project   b, d, m

9. Chipko movement   a, c, i, j, k

Unit 7

Multiple Choice

1. b. it was overhunted even though its habitat remained intact.

2. c. their breeding must be managed to reduce inbreeding.

3. a. the complicated structure of the forest gives them many places to live.

4. d. it depends on undisturbed rain forest for its survival.

5. a. they needed those resources in years that their crops failed.

6. a. to give villagers alternative livelihoods that reduced their need to invade the park.

7. c. the lake food chain can be manipulated to support a bigger catch.

8. b. depends on a water source that is also diverted for human uses.

9. b. has brought back a tree species that allows other species to recover.

10. a. consumes the basic production of the ecosystem just as a large animal might.

11. c. reducing military spending is one source of money needed for conservation.

Unit 8

Multiple Choice

1. c. converting industrial boilers to burn straw from nearby farms.

2. b. the same amount of light can be produced with far less energy than that needed by a conventional bulb.

3. b. Volvo's management felt that customers were not interested in fuel efficiency.

4. e. (a, b)

   a. smoke from stoves threatens the health of Indian women.

   b. firewood shortages are common in India.

5. c. suggests what energy sources may look like as humanity makes the transition to reliance on renewable sources.

6. e. (a, c)

   a. electricity generation is not the only use of fossil fuels that releases carbon dioxide.

   c. important questions of safety, waste disposal, and public acceptability have not yet been solved.

Fill In

1. A Danish system that combines windmills and manure-fired biogas to produce electricity and heat is called (LOCUS). This is one form of a more general arrangement called (cogeneration) in which otherwise wasted energy is tapped to provide needed energy services.

2. Brazil uses sugar cane to produce (alcohol) and (biogas) as alternative energy sources. The first is a substitute for (gasoline), whereas the second is used to generate (electricity).

3. The pressurized fluidized-bed technology shown in the program depends on burning (coal). It reduces the emissions of (sulfur dioxide) and (nitrogen oxides) that cause acid rain, but like all technologies that rely on fossil fuels, it releases (carbon dioxide).

Matching

1. pressurized fluidized-bed
    coal combustion   b, d, e, g, h

2. high-temperature
    gas-cooled nuclear
    fission reactor   b, c, e, f, g

3. improved wood stoves   a, d, i

4. compact fluorescent bulb a, g, h

5. photovoltaic cells   a, e, h

6. biogas generators   a, h

7. windmills   a, e, g

8. cogeneration plants   a, b, d, g

9. LCP 2000   a, b, d, g

Unit 9

Multiple Choice

1. c. The main source of plant nutrients is
        purchased off the farm.

2. b. the opportunity to sell U.S. grain to buyers
        overseas.

3. d. only a small amount of the chemicals
        sprayed actually reaches the target crop.

4. c. suspicious but inconclusive because the
        scientific data are incomplete.

5. a. a combination of population pressure and
        economic forces is leading farmers to
        specialize in cash crops and reduce the
        diversity of the gardens.

6. d. a major pest of rice developed resistance to
        widely used pesticides and threatened the
        country's harvest.

7. e. agroforestry methods.

8. e. (a, c, d)

    a. the region's soil was highly fertile and
        erosion seemed harmless.

    c. uncontrolled water runoff led to creation of
        severe gullies.

    d. opportunities to export wheat led farmers
        to emphasize production.

9. e. (b, c)

    b. construct checkdams and other water-
        harvesting structures.

    c. learn better management of the internal
        resources available to them.

10. c. farms using low-input approaches can be
         both productive and profitable.

11. c. could be achieved if money is available for
         research on yields.

True or False

1. The green revolution is now practiced by most
    farmers in developing countries.   (false)

2. Problems with resistance to pesticides, though
    common in developing countries, rarely occur
    in countries like the United States.   (false)

3. Integrated pest management sometimes uses
    chemical methods.   (true)

4. Unlike Australia, the United States solved its
    erosion problems after the Dust Bowl
    experience.   (false)

5. The greatest problem facing farmers in the
    Sahel region is water management.   (true)

6. The risks of exposure to agrochemicals are well
    understood.   (false)

Unit 10

Multiple Choice

1. e. (b, d)

   b. plastics account for a growing share of American trash.

   d. packaging makes up the bulk of American garbage.

2. e. public attitudes are a key to acceptance of waste management facilities.

3. b. save money by recovering valuable materials used in manufacturing circuit boards.

4. c. a cost-effective method of controlling waterborne diseases.

5. b. biological treatment of human waste can complement conventional sewage treatment.

6. b. reduces waste volume by three-quarters.

7. d. are renewable in principle but can be depleted in practice.

8. b. in landfills, surface lagoons, and injection wells.

9. b. generates sludge that must also be disposed.

True or False

1. Recycling and incineration together account for more than half the solid waste disposed in the United States.   (false)

2. Japan incinerates about 3% of its solid waste. (false)

3. The United States incinerates about 3% of its solid waste.   (true)

4. Paper recycling is more widespread in the United States than in Japan.   (false)

5. Japan's limited size makes landfills a disposal method of last resort.   (true)

Unit 11

Multiple Choice

1. e. (a, c)

   a. the country's birth rate is one of the highest in the world.

   c. the present population is near the limits of the country's productive land.

2. d. direct payments to families who have no more than two children.

3. b. allowing some agricultural settlement and encouraging commercial activities based on wildlife.

4. b. wildlife is a prime tourist attraction that brings money into the country.

5. d. about half as high as the rate in Zimbabwe.

6. c. life can be better for families that limit their number of children.

7. c. using education to supplement limited enforcement activities.

8. e. (a, b, c)

   a. timber companies that ignored replanting rules.

   b. a growing population's need for farmland.

   c. government emphasis on agricultural development at the expense of forests.

9. a. providing rural communities with alternatives to forest destruction.

10. c. beginning to focus government attention on long-neglected problems.

11. e.   (b, d)

    b.   a population that is nearly stable.

    d.   almost none of its original native forests left.

12. a.   (b, d)

    b.   waste from a detergent-manufacturing plant.

    d.   sewage from households in the city.

13. e.   (a, c)

    a.   forest practices involve clearcutting, which exposes large areas to erosion.

    c.   nearly all the country's natural forests have been converted to single-species plantings.

Unit 12

Multiple Choice

1.   b.   the removal of trees for firewood and farmland.

2.   c.   most Kenyan farmers are women and soil erosion affects them directly.

3.   a.   poverty is both the cause and the effect of environmental deterioration in many developing countries.

4.   c.   the lake was considered a national treasure because of its pristine quality.

5.   d.   toxic air pollutants released by routine activities of residents near the lake.

6.   e.   have organized and challenged local authorities to take action against polluters.

7.   d.   was one of the world's first efforts to control pollution on a regional basis.

8.   e.   (b, c, d)

    b.   collaboration on monitoring helps build trust among the countries involved.

    c.   monitoring provides baseline data used to determine appropriate responses.

    d.   scientists could help interest political leaders in taking action on pollution.

9.   c.   is being addressed by an unusual effort to recover waste oil for reuse.

10.   a.   collection of scientific data and a framework for international cooperation.

11.   b.   they could not afford the more costly substitutes for CFCs.

12.   e.   (a, b, d)

    a.   using funds formerly spent by the military to finance sustainable development.

    b.   shifting leaders' attention from military issues to broader questions of quality of life.

    d.   reducing suspicions and increasing cooperation among societies.

True or False

1.   About 120 nations have now signed the Montreal Ozone Agreement to cut back on CFC production.   <u>(false)</u>

2.   Scientific evidence suggests that the CFC cutbacks proposed in the Montreal Ozone Agreement will not go far enough to protect the ozone layer.   <u>(true)</u>

3.   Greenpeace and Friends of the Earth believe CFCs can be used safely in small quantities.   <u>(false)</u>

4.   The British approach involves voluntary rather than mandatory controls on CFC producers.   <u>(true)</u>

5.   The inability of nations to agree on the CFC issue makes progress on other atmospheric problems unlikely.   <u>(false)</u>